Urban Sustainability

The Urban Sustainability Book Series is a valuable resource for sustainability and urban-related education and research. It offers an inter-disciplinary platform covering all four areas of practice, policy, education, research, and their nexus. The publications in this series are related to critical areas of sustainability, urban studies, planning, and urban geography.

This book series aims to put together cutting-edge research findings linked to the overarching field of urban sustainability. The scope and nature of the topic are broad and interdisciplinary and bring together various associated disciplines from sustainable development, environmental sciences, urbanism, etc. With many advanced research findings in the field, there is a need to put together various discussions and contributions on specific sustainability fields, covering a good range of topics on sustainable development, sustainable urbanism, and urban sustainability. Despite the broad range of issues, we note the importance of practical and policy-oriented directions, extending the literature and directions and pathways towards achieving urban sustainability.

The series will appeal to urbanists, geographers, planners, engineers, architects, governmental authorities, policymakers, researchers of all levels, and to all of those interested in a wide-ranging overview of urban sustainability and its associated fields. The series includes monographs and edited volumes, covering a range of topics under the urban sustainability topic, which can also be used for teaching materials.

Ali Cheshmehzangi · Tian Tang

30 Years of Urban Change in China's 10 Core Cities

Springer

Ali Cheshmehzangi
School of Architecture, Design and Planning
University of Queensland
Brisbane, QLD, Australia

Tian Tang
Independent Researcher
Chongqing, Sichuan, China

ISSN 2731-6483 ISSN 2731-6491 (electronic)
Urban Sustainability
ISBN 978-981-97-8845-3 ISBN 978-981-97-8846-0 (eBook)
https://doi.org/10.1007/978-981-97-8846-0

This Springer imprint is published by the registered company Springer Nature Singapore Pte Ltd.
The registered company address is: 152 Beach Road, #21-01/04 Gateway East, Singapore 189721, Singapore

Photo taken by Tian Tang in Beijing, June 2021

It was all fine until you spoke out the truth!

With some doubts and extra thinking, we decided to dedicate this book to those radical (or better to say political) snakes with little minds; those who have no innovation but to belittle the greater minds and souls. Their lack of vision follows their underprivileged mindsets, which ultimately represent their long-lasting obsessions, pretentious attitudes, and obnoxious acts—May they grow, perhaps to acquire some humanity before the end of their times.

Prologue

> *"Success should always call for showing greater kindness, generosity, and justice; only people lost in the darkness treat it as an occasion for greater greed"*
>
> —Cyrus the Great, the first Persian King

China's urban landscape has undergone a remarkable transformation over the past three decades, boosted by unprecedented economic growth, rapid urbanisation, and ambitious government policies. The country has experienced major changes and shifts in its development, but also has gone through major transformations, policy reforms, and novel transitions. From bustling metropolises to newly emerging urban centers and new development zones, China's urban development has become a symbol of its ascent on the global stage. Many countries, particularly in the context of developing world, now look up to China as a role model in growth and development. China's narrative of urban change encapsulates not only the physical expansion of cities but also the socio-economic shifts and environmental challenges that accompany such rapid growth. At the heart of China's urbanisation lies the convergence of millions of individuals seeking better economic prospects in the cities. This is not new to China and has happened in many other countries before, too. This influx of rural migrants has not only reshaped the demographic fabric of urban areas but has also spurred significant socio-economic changes. Again, this has happened elsewhere, but the scale of China's rural-to-urban migration is just the major difference we see in the country's new era of transformation. Rural-to-urban migration has led to the formation of vast urban labour pools, driving industrialisation, and fueling the growth of manufacturing hubs. We should not neglect the fact that such rapid process helped China take 300 million people out of poverty in less than 30 years. Without a doubt, this journey has been incredible, but it also indicates the scale of change that helped a country become what it is now. More importantly, China's rural-to-urban migration has also facilitated the rise of a new urban middle class, characterised by increased consumption patterns, higher living standards, and aspirations for upward mobility. We talk about these issues to our students, asking them to discuss with their parents and grandparents, comparing the living standards of now with a few decades ago. The conclusion of those discussions is that not many countries have ever reached what China has in a such short period of time in our recent history.

The period from the late 20th century to the present has witnessed an supreme wave of urbanisation in China. This phenomenon has been primarily driven by rural-to-urban migration as millions of people sought better economic opportunities in the cities. Government-led reforms, including the relaxation of residency restrictions and the promotion of industrialisation, spurred this mass migration, leading to the formation of mega-cities like Beijing, Shanghai, and Guangzhou.

However, alongside the so-called 'greater opportunities' presented by urbanisation, there are always significant socio-economic challenges to contend with. These should not be neglected and nor should they be undermined in the next steps of growth and development. Income inequality has become increasingly pronounced, with disparities between urban and rural areas widening and income gaps within cities deepening. While some have prospered in the urban landscape, others, particularly migrant workers and marginalised communities, continue to face barriers to accessing adequate housing, education, and healthcare. Many agencies and stakeholders are fully aware of the conditions particularly that sacrifices had to be made before in order to reach where we have reached today. This stark urban-rural divide underscores the complexity of China's urbanisation journey and highlights the need for inclusive development policies that address the needs of all citizens. Nonetheless, China's attention to regional development and high-quality urbanisation means a new era of development beyond just the existing business-as-usual scenarios. Also, the growing position or status of China at the global level means China is and will be playing a more leadership role. This is not only China's intention for its global leadership role, but also many countries' expectation of China.

One of the defining features of China's urban development has been the construction of iconic skyscrapers and modern infrastructure. Many of our friends in their visit to China describe the country's modern infrastructure to be unique and remarkable. Yet, in parallel to these appraisals and commendations, we question what development means to people? How should development be perceived in the future? And what new directions could separate China's development path from the rest of the world? We highlight the fact that cities have been transformed into sprawling hubs of commerce, culture, and innovation, with futuristic skylines dominated by towering buildings. This rapid vertical and horizontal expansion has not only reshaped cityscapes but also symbolised China's ambition to become a global economic powerhouse. But as Nelson Mandela puts it well, "*after climbing a great hill, one can only finds that there are many more hills to climb*". Thus, there is no clear end to current flows and progressions unless long-terms goals start emerging to make progress more defined, humanistic, and reflective on what the future will bring to the nation and beyond.

Alongside the physical transformation, China's urbanisation has brought about profound social and economic changes. China's narrative of urban change is shaped by ongoing efforts to harness the potential of technology and innovation in urban governance and management. Many development and city branding methods have gained traction as means of enhancing efficiency, improving public services, and fostering economic development through the integration of digital technologies and data-driven solutions. From smart transportation systems to intelligent infrastructure and digital governance platforms, Chinese cities are embracing innovation to tackle urban challenges and enhance the quality of life for residents. The influx of rural migrants has fueled the growth of a new urban middle class, driving consumerism and creating demand for services ranging from luxury goods to high-end healthcare.

Furthermore, China's urbanisation has placed immense pressure on the environment and natural resources. The rapid expansion of cities has led to pollution, congestion, and the depletion of water and energy resources. In response, the Chinese government has implemented various initiatives aimed at promoting sustainable urban development, including investments in green technologies, the development of eco-friendly infrastructure, and the promotion of public transportation. Leveraging advancements in technology such as artificial intelligence, big data, and the Internet of Things, Chinese cities are striving to improve efficiency, enhance urban governance, and provide better services to residents. These smart city initiatives encompass various aspects of urban life, including transportation, healthcare, education, and public safety.

It is important to note that China's urban development has not been limited to its megacities or larger economic hubs across the country. Secondary and even tertiary cities have also experienced significant growth and transformation, fueled by government policies aimed at decentralising economic development and alleviating pressure on major urban centers. Such shift was fueled through China's new-type urbanisation plan that has changed the structure and dynamics of development at the regional and national levels. This decentralisation strategy has led to the emergence of new urban clusters and regional hubs, contributing to a more balanced distribution of resources and opportunities across the country.

Looking ahead, China's urbanisation is expected to continue unabated, albeit at a more moderate pace. The government's commitment to urban development remains steadfast, with ambitious plans outlined in initiatives such as the Belt and Road Initiative (BRI), the Xiong'an New Area project, the Silicon Valley of the East, the PuDong of the West, etc. However, as China enters a new phase of urbanisation characterised by quality over quantity, the focus is shifting towards addressing issues of sustainability, inclusivity, and resilience in order to build cities that are not only prosperous but also livable for all residents. Hence, this book provides us a great understanding of micro-history of China's recent development, particularly aligned with sustainable development plans of its major regions, core cities, and urban hubs. We hope this becomes an insightful book for those who genuinely seek for sustainable development from multiple perspectives. In essence, China's urbanisation story is not just about the physical transformation of cities but also about the intricate interplay of socio-economic dynamics and environmental considerations. Hence, there is more to explore, there is more to do/see/observe/suggest/plan, and there is more to read.

Brisbane, Australia Ali Cheshmehzangi
Chongqing, China Tian Tang

Acknowledgements

This book is the signature output of our joint efforts between the glorious but challenging years of 2022 to 2024. We thank our industry partners, local governmental agencies, and local governments of various levels in China for their continuous support while we were conducting research and completing case study analysis. We held several major industry workshops, which were helpful for us to have a better picture of recent urban changes and transformations in China. We have been enthusiastic in researching these important topics and will continue working strategically and healthily with key stakeholders to achieve higher goals.

There were once mischievous and disruptive attempts to sabotage a fast-moving train, not knowing the next stops were redemption and a sort of destiny we had in mind down the line. Destiny, we say, but a pathway that we did not envisage as such at all. Written in this life, perhaps; and to be restarted again and again, we are yet to see. Nonetheless, we have not settled to halt our progressive mindsets, and the so-called inheritance shall remain for years to come. They, who tried, are seeing the abundant fruits from longer distances.

To acquire opportunities and thriving ongoing manifestations, younger occurrences unify failures and innovative lightbulb-effects. Education drives young opportunists, unfolding loyalty, improvements, trust, talents, liberty, and equalities. So hard it tried scantily—so hard it tried, for attracting creative eloquences slightly. Most of the happenings enunciated reproduction of failing unjust crooks while killing enthusiastic risings so swiftly. This propagation can only be seen (if still concealed reflexively) with an open eye or through the complexity of words of thoughts and their conjuring agglomeration chained at the first stage of each stop. The same that we witnessed in vivid—yet enigmatic—notes and writings in the midst of an island breeze at the midnight of spring in mid-2021. And equivalently the same (but not similar) that we have seen in unceasing efforts of making the visible known through unforeseen knowledge and supplemented communications. Time has shown that, and it has shown us a lot.

Moreover, we acknowledge contributions and full support from our team members in three locations, who helped complete related case study work with us. We thank those who attended our workshops, webinars, and report review sessions. We thank

Miss Shi Yao for her support in collecting materials and tidying things up towards the end of this exciting project. It is confirmed that certain assisted tools have been used in this data gathering and preparation. She and the Little Prince have a few things in common, but she still does not know that. Lastly, we continue to thank our families and genuine friends, near and far. Only those who endured throughout privations are cherished as our authentic chums.

About This Book

As a continuation of our award-winning book '*China's City Cluster Development in the Race to Carbon Neutrality*', this book covers China's major urban changes over the last 30 years. Unlike the previous book, where we highlighted regional development issues, this book explicitly explores the city cases, particularly those that are considered core cities in China. Based on the micro-historical analysis of China's urbanization trend and urban development patterns, we see that cities in China have played a significant role in driving the country's sustainable development agenda. In a way, they have had both positive and negative impacts on achieving sustainable development. We look at these last three decades mainly because, during this period, China's urbanization became unprecedented, the central government made several pledges and signed many international agreements related to sustainable development, and China grew rapidly to become the second global economic power. Aligned with the overarching Belt and Road Initiative (BRI) and the more recent Carbon Neutrality Plan (CNP), Chinese core cities played a significant part in regional development, urban-rural relations, industrial clustering, the development of free trade zones and special economic zones, etc. All these recent developments are due to China's ongoing urbanization and urban development.

'30 Years of Urban Change in China's 10 Core Cities' is a mapping study of China's core cities and their changes over three decades of rapid urbanization, urban growth, and economic development. We explicitly highlight each selected city's development trends, sustainable development plans, and major strategies. In this Volume (out of our two 'connected' volumes), we summarise lessons learnt from all 10 case study examples, and we hope they can be utilized for other developing and rapidly urbanizing nations. We expect this book to be a valuable resource for local governments, authorities, urban planners, urbanists, practitioners, developers, and urban researchers. We trust China is no longer a developing country, and much of these recent progressions are owed to its structured urban development strategies, robust governmental structure, and progressive attitude to growth and development. Hence, this topic coverage at the point where China's urbanization is shifting to high-quality urbanization is essential and beneficial to multiple stakeholders.

Here, we provide a map to highlight the location of all 10 core cities studied in this first volume of the two books covering the topic. This is intended for non-Chinese scholars or those who may not be familiar with the context of China. These cities are ordered in the following:

1. Beijing Municipality, The Capital of China, a direct-administrated municipality
2. Shanghai Municipality, a direct-administrated municipality
3. Shenzhen City, Guangdong Province
4. Guangzhou, Guangdong Province
5. Hangzhou, Zhejiang Province
6. Chengdu, Sichuan Province
7. Nanjing, Jiangsu Province
8. Wuhan, Hubei Province
9. Tianjin Municipality, a direct-administrated municipality
10. Suzhou, Jiangsu Province

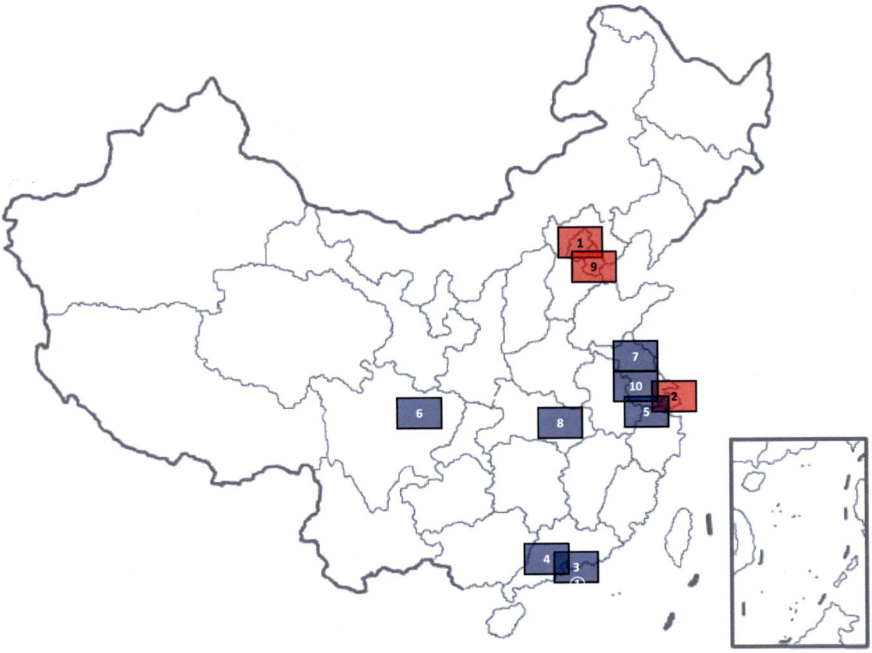

Contents

About the Authors

Ali Cheshmehzangi is a Full Professor and Head of the School of Architecture, Design and Planning (ADP) at the University of Queensland, where he leads the school and researches across the fields of architecture, design, and planning/urbanism. He has been the World's top 2% field leader for two consecutive years, recognised by Stanford University. He is among the top 30 global scholars in the urban sustainability research area. Ali is internationally known for his scholarly contribution and extended work on climate resilience and sustainable urbanism research. Ali has a wealth of experience, expertise, and a strong commitment to innovation and excellence in the fields of architecture, design, and planning, the three core disciplines of his School at UQ. With a career spanning over two decades, he has made significant contributions to the academic and professional communities, with a focus on sustainable and environmentally conscious design. Prior to joining UQ, Ali held several strategic leadership and senior managerial roles, such as Vice-President for International Engagement and Global Partnership, Founding Director/Head of the Center for Innovation in Education and Research, Head of the Department of Architecture and Built Environment, Founding Director of the Urban Innovation Lab, Director of a university-wide Teaching and Learning platform, Director of International Research Network for Rural and Urban Development, Head of Research Group for Sustainable Built Environment, co-director of university-wide research priority areas, Director of Center for Sustainable Energy Technologies, and Interim Director of Digital Design Lab. So far, Ali has published over 500 journal papers, articles, conference papers, book chapters, and reports. He also has 33 other academic books, three of which have received awards at the national, provincial, and municipal levels. He also has received international awards and recognition for his research on urban resilience studies and sustainability research, as well as a Vice-Chancellor's award for his impactful contribution to higher education.

Tian Tang is an independent researcher focused on analysing China's urban and regional transitions and urban sustainability research. She co-authored a book with Ali Cheshmehzangi, titled "*China's City Cluster Development in the Race to Carbon Neutrality*" (2022), which received a prestigious national award from Springer in

November 2023. For almost a decade, Tian has worked with several local agencies and consultancy firms, with significant research work in Sichuan, Shanghai, Guangdong, Hainan, Hunan, and Chongqing, all in China. Tian is experienced with fieldwork activities in China and currently focuses on analysing urban transitions in urban clusters and specific regions in China, while exploring global case study examples, too. With extensive collaborations with the industry, relevant institutes, and governmental agencies, Tian has developed an extensive profile in urban sustainability research. In 2016 and then from 2019 onwards, her extended work included several case study research works as an independent research associate member and industry collaborator. In 2020-22, she focused on carbon neutrality research, case study reports, regional analysis, Chinese city clusters, and ICT-driven smart-resilient urban transformations. To date, Tian has 120+ reports, papers, and conference/workshop presentations and is co-author of an award-winning book.

Chapter 1
China's Urbanization Progress and Urban Change: Sustainability Cannot Be Divorced from China's Future Development

1 A Brief Introduction to China's Urbanisation Progress and Urban Change

Undoubtedly, China's urbanisation progress is one of the most remarkable transformations in modern history. It is a story of unprecedented speed and scale, driven by both policy decisions and economic forces [19, 50, 51, 72, 82]. From the historical context, we see China has a long history of urbanisation and even a much higher rate before compared to most developed countries. But, China's modern urbanisation can be traced back to the late 1970s when economic reforms, notably the opening up policies led by Deng Xiaoping in 1978 (i.e., the Open Door Policy), spurred rapid industrialisation and urban growth [13, 15, 17, 41, 60]. This new era led China to open for new directions and investment, focusing on economic reforms, industrialisation drive, urban growth and migration, foreign investment, technology transfer, and social and cultural changes [31, 52, 90, 94]. The multiplicity of this opening up was so significant that created a foundation for China's emergence [7, 18, 33] as a global economic power later on.

Through this stage of rapid urbanisation, we see unprecedented growth and rural-to-urban migration mainly driven by China's focus on economic development and growth. Since the late 1970s, China's urban population has grown exponentially, highlighting the fact that we always argue that 'there is no non-urbanised developed country in the world'. Knowing the importance of urbanisation, China reinitiated its economic engines across the country, focusing primarily on larger cities and mainly on coastal cities of east and south. Millions of people have migrated from rural areas to cities in search of better economic opportunities, leading to the rise of megacities like Beijing, Shanghai, and Guangzhou [28]. This never happened before in the context of China, and hence, government policies started pouring in through various mechanisms, new platforms, and institutions [12, 22, 57, 92]. The Chinese government has actively promoted urbanisation through various policies, such as

A. Cheshmehzangi and T. Tang, *30 Years of Urban Change in China's 10 Core Cities*, Urban Sustainability, https://doi.org/10.1007/978-981-97-8846-0_1

the "Hukou" system reform [21, 29], which allowed rural migrants to obtain urban residency permits, granting them access to urban services and benefits.

Since 2014, strategic changes through the new-type urbanisation plan started a new era of steady but high-quality urbanisation [76]. This has much benefitted for years from extensive infrastructure development, where China has invested heavily in infrastructure to support its urbanisation efforts [83]. There are great success stories as a result of this investment, one of which is the excellent high-speed railway network across the country. Such infrastructure development plans include the construction of highways, high-speed rail networks, airports, and urban public transportation systems, facilitating connectivity within and between cities. The backbone of all these is China's gradual but effective upgrades in urban planning and development. Evidently, urban planning has played a crucial role in shaping China's cities. The government has implemented master plans to manage urban growth, control land use, and address environmental concerns. Some examples were tested in pilot zones or new development areas, and some have been scaled up in recent years—e.g., some through policy experimentation [38]. However, rapid urbanisation has also led to challenges such as housing affordability, traffic congestion, and environmental pollution, which are commonly associated with larger cities in China.

When we refer to China's urbanisation, we simultaneously talk about economic transformation(s) that are embedded in the country's economic restructuring and ambitious plans. It is evident that urbanisation has been closely tied to China's economic transformation [11, 17, 37]. Cities have become centers of industry, commerce, and innovation, driving economic growth and prosperity. There are many excellent examples where smaller new zones or existing clusters of townships become successful cities or economic hubs [88]. At the same time, there are also examples of decay in places where economic development did not pick up as planned. In all cases, the growth of China's middle class has also fuelled consumer demand, further stimulating urban development [5, 17, 23, 35, 53, 73]. This increase in supply and demand has motivates inward and outward looking strategies towards local, regional, and national economic development.

Aligned with China's plan to eradicate poverty and ensure it develops towards a prosperous country for the majority, we have witnessed major social changes across the whole country. Existing literature and case study reports indicate the fact that urbanisation has brought about significant social changes in China. The changes in localities, socio-cultural attributes, and social political systems of many regions meant new opportunities for social changes that seem essential for China's next stage of development [70]. These social changes, particularly in recent few decades, led to increased social mobility, as rural migrants have found employment opportunities in cities. As China targets a much higher urbanisation rate at the end of this decade (i.e., estimated to reach 70% of urbanised population by 2030), we still witness changes that take place in small and medium sized cities. Aligned with China's new-urbanisation plan, we see greater opportunities for the development of enhanced social infrastructure and social capacities in particular regions [19, 56, 79, 83]. Perhaps, much of the focus has been on larger cities in recent decades, but we foresee new opportunities in the horizon through new development zones [36] and a more comprehensive focus or consideration of regional development at large. Despite its achievements, China faces numerous challenges in sustaining its urbanisation process. These include managing urban sprawl, addressing environmental degradation, providing affordable housing, and ensuring social integration. Moving forward, China is likely to focus on promoting sustainable urban development and improving the quality of life for its urban residents. At the same time, the country needs to focus on sustaining its food security, social security, and environmental protection plans, which are the backbone of regional development plans. Nevertheless, we could argue that China's urbanization progress is a complex phenomenon driven by a combination of economic, social, and political factors [54]. This may be reminiscent of everywhere else that experiences growth and development, but the scale and pace of development in China is simply a unique example around the globe [4, 30, 49]. While it has brought about unprecedented growth and development [26], it also poses significant challenges that require careful planning and management.

This ice-breaking introduction chapter briefly highlights what China's urbanisation and its progress mean to us, particularly that we aim to delve into smaller scale of cities where the actual development is taking place. China's urbanisation is unprecedented due to the sheer magnitude of its population. China's demographic factor of more than 1.4 billion people magnifies the scale of urbanisation efforts [42]. The need to accommodate and provide for such a vast number of people presents immense challenges in terms of infrastructure development, housing provision, and social services. Nonetheless, cities have gone through transformative changes over these recent decades just to ensure they can support the country's economic growth and development. Unlike many other countries where urbanisation often occurs organically or in response to market forces, China's urbanisation has been heavily influenced by government policies and planning. The Chinese government plays a central role in directing urbanisation efforts, guiding investment, and managing land use. While this top-down approach has facilitated rapid development, it has also led to issues such as inefficient land use, urban sprawl, and overdevelopment in some areas. But in most cases, cities have grown rapidly to form into economic hubs and main engines of China's growth-led strategies. Thus, we see higher demand for infrastructure while the growing environmental concerns seem endless.

This book is a continuation of what we covered before at the regional level. The main difference is scale, where we focus on city-level interventions and directions rather than smaller scale or larger level plans. Prior to this book, we have covered China's sustainable urbanism at neighbourhood or community level [18] and also delved into regional development opportunities for sustainable development progression and carbon neutrality plans [19] that are expected to influence most (if not all) of the development strategies in the next few decades. Looking ahead, China's urbanisation is anticipated to prioritise sustainability and resilience from multiple perspectives or even multiple directions. There should be no one solution to growing concerns about socio-cultural and socio-political changes as well as environmental concerns that seem to very much tangled with China's pledge to high quality and low carbon development. This includes promoting green urban development, enhancing public transportation, improving energy efficiency, and addressing social equity issues. In this regard, achieving sustainable urbanisation requires a holistic approach that balances economic growth with environmental protection and social justice.

In the next two sub-sections, we provide a summary of a micro-history analysis or mapping of China's development. In fact, the whole book should be read and understood as a sort of mapping study, connecting the dots between various aspect of city-level/urban-level development in China. By 'micro-history' we merely refer to the last 30 years, where China's urbanisation rate became double digit at some point and then stabilised to a rapid age of economic growth and development that led to the current age of high-quality urbanisation ideology or direction. Afterwards, we provide a summary of China's recent glorious past aligned with ongoing pathways of dual carbon neutrality plan, including decarbonisation strategies and low carbon development plans that feed into China's sustainable development targets. This discourse would then align well to our further reflection on China and the Sustainable Development Goals (SDGs) before we explore the movements from slogans or general propaganda and towards the reality of implementing transformative changes at multiple scales. We then provide a summary of what lies ahead for cities, regions, and new development zones in China before highlighting aim and objectives of the book. Lastly, as the book title entails, we present our comprehensive selection of 10 core cities, a narrative that we believe helped us understand what could be defined and recognized as a core city in China. Through this selection process, we introduce different understanding of considerations and evaluations before we close the chapter with a reflective overview.

1.1 A Micro-history Analysis of China's Development: The Era of Rapid Urbanisation and Growth

China's development over the past three to four decades has been nothing short of remarkable, especially in terms of urbanisation and economic growth. This brief micro-history analysis delves into the intricacies of China's transformation during this period, focusing on key events, policies, and trends that have shaped its urban landscape and drove its economic expansion. This relatively short period is a testament to the resilience, adaptability, and strategic vision of its leadership and people. It is an era that we refer to, in many events, as the time China went through rapid learning curves and developed at least three times faster than developed nations. This period has witnessed a profound transformation, particularly in terms of urbanisation and economic growth, putting China in a rather uncomfortable position of leadership alongside other nations, including the United States [19, 47, 65, 91]. The "two global leadership" mindset is still a new ideology, but remains as a valid discourse (or even a strategic plan) related to how China has been and is positioning itself as a leader. Such transformative era of development has, to some extent, driven China onto the global stage as a major economic powerhouse, now second only to the US. A closer examination of this transformative journey reveals a tapestry woven with intricate threads of policies, events, and trends that have shaped China's urban landscape and driven its economic expansion. Here, we provide some of the critical timelines of these recent few decades.

1.1.1 1980s: Opening Up and Early Reforms

The late 1970s marked the beginning of China's economic reforms under Deng Xiaoping, ushering in an era of opening up and modernisation [19, 40, 58, 68, 78]. This happened right after the end of a decade-long Cultural Revolution, setting a mark for immediate changes in how China would develop in the years after. In 1980, the establishment of Special Economic Zones (SEZs) such as Shenzhen and Zhuhai instigated the influx of foreign investment and the development of export-oriented industries [6, 45, 55]. Such process helped revamping some of the traditional setups [20, 80, 81], particularly around regional development, and industrial structures. This period witnessed a gradual shift of population from rural to urban areas, laying the groundwork for future urbanisation trends. The creation of SEZs not only accelerated the pace of urbanisation but also contributed to the transformation of China's urban landscape. Formerly small fishing villages like Shenzhen were rapidly transformed into bustling metropolises with skyscrapers, factories, and modern infrastructure. This transformation was accompanied by the development of supporting industries such as manufacturing, trade, and logistics, further fuelling urban growth and attracting additional migrants from rural areas. In addition to the establishment of SEZs, other economic reforms introduced during this period, such as the de-collectivisation of agriculture and the liberalisation of trade and investment policies, also played a significant role in driving urbanisation [59, 62, 64, 89]. As rural households gained greater autonomy over their land and economic activities, many opted to migrate to urban areas in search of non-agricultural employment opportunities. This internal migration contributed to the swelling ranks of urban residents and the expansion of China's urban population. This era is also known for the starting point of further investment and expansion in infrastructure development. It is indeed a turning point in the country's urbanisation rends, characterized primarily by rapid industrialisation, urban expansion, and population migration from rural to urban areas.

1.1.2 1990s: Accelerated Urbanization and Infrastructure Boom

The 1990s saw a significant acceleration in China's urbanisation process, fuelled by rapid industrialisation and urban expansion [34, 77]. This is the time when China's urbanisation rate was double digit, which created a massive transformative change across the country. The urban landscape was already changing through new investment opportunities, and China took a faster pace in developing the physical infrastructure that was crucial to the country [10]. This so-call infrastructure boom was the result of China's long-term goals about rapid industrialisation and development which were already affected by earlier economic reforms. The necessary readjustments in governmental institutions meant new opportunities beyond initial reforms and towards transformative changes and restructuring that helped boosting the country's development [67]. For instance, in 1994, the government launched the "Develop the West" campaign, aiming to boost economic development in China's western regions [15, 32, 39, 66]. This initiative led to the construction of infrastructure projects such as highways, railways, and urban utilities, facilitating the migration of rural populations to newly developed urban centers.

1.1.3 2000s: Rise of Megacities and Economic Expansion

The early 2000s witnessed the rise of megacities like Beijing, Shanghai, and Guangzhou as economic powerhouses driving China's growth. In 2001, China joined the World Trade Organization (WTO), further integrating its economy into the global market and attracting foreign investment [1, 27]. This period also saw the implementation of urbanisation policies aimed at managing the rapid influx of migrants [71] and addressing urban–rural disparities through initiatives like the "New Socialist Countryside" campaign [2]. This campaign aimed to modernise rural areas, improve infrastructure, and provide better living conditions for rural residents, thereby reducing the pressure on urban centers and fostering balanced regional development. The implementation of such policies highlighted China's commitment to sustainable development and social cohesion amidst rapid urban expansion. The influx of foreign capital, coupled with domestic economic reforms, fuelled rapid urbanisation and industrialisation across the country. Beijing, Shanghai, and Guangzhou emerged as epicenters of this economic boom, attracting businesses, talent, and investment from around the world [46, 93]. These megacities not only served as engines of economic growth but also as centers of innovation, culture, and modernisation.

1.1.4 2010s: Sustainable Development and Urban Renewal

The 2010s marked a shift towards sustainable development and urban renewal as China grappled with environmental challenges and social inequalities. In 2013, President Xi Jinping announced the ambitious Belt and Road Initiative (BRI), aiming to enhance regional connectivity and infrastructure development across Asia, Africa, and Europe [69]. Domestically, policies promoting green urbanisation and eco-friendly initiatives gained prominence, alongside efforts to upgrade urban infrastructure and improve living standards in both megacities and smaller urban centers. This decade was one of the most prominent times for China's progress in setting up new platforms, policies, directions in sustainable development. Following the earlier work on ecological civilization movement in 2007, China's approach became more explicit and more associated with its plan for further development and urbanisation strategies [61]. During the 2010s, China underwent a significant transformation in its approach to sustainable development and urban renewal, driven by rapid urbanization, environmental challenges, and the need for more efficient resource utilisation. This meant some of the primary regions or provinces had to start rethinking their strategies to meet the national agenda and guidelines. This decade is also a critical time for China due to its emerging and continuing strategies aligned with the SDGs, Paris Agreement, low-carbon development, and many pilot projects that went viral 2009 onwards [24]. Several key aspects define China's development in these areas during this decade include transformative development in economic growth, technological advancement, global influence, and social changes that seemed inevitable through the rapid pace of urbanisation. Under technological advancement, for instance, China made significant strides in technological innovation in this decade. It only a few years, it became a global leader in various fields, including telecommunications, artificial intelligence (AI), and renewable energy. For the latter, as an example, China became the leading country to have the largest renewable energy capacity. High-tech companies like Huawei, Tencent, Alibaba, and Xiaomi emerged as major players in the global technology industry [25], which again aligned very well with China's smart development [43], digital-driven management [14], and the recreation of platforms that started to develop integration between different governmental units and agencies.

1.1.5 2020s: From Low-Carbon Development to Carbon Neutrality Plan

In the 2020s, despite immediate disruptions of the COVID-19 pandemic, China embarked on a transformative journey from prioritising low-carbon development to announcing and implementing a comprehensive Carbon Neutrality Plan (CNP) in October 2020. Soon after the direction shifted drastically towards meeting the dual carbon targets or strategy, i.e., achieving carbon peak by 2030 and reach carbon neutrality before 2060 [86]. This shift marked a significant milestone in China's efforts to address climate change and transition towards a more sustainable and green economy. Immediate shifts were witnessed in low-carbon development initiatives, revamping some of the earlier regional development plans [19]. This meant China started building upon its previous efforts to mitigate carbon emissions, which then brought the vision to intensify its focus on low-carbon development in the early 2020s. Although embedded in the guidelines and policies of the 11th Five-Year-Plan (FYP) onwards (if not earlier), low carbon transitions were quickly redefined into transformative changes that seem essential to achieving the 2030 and 2060 targets [17]. This no-so-new development mindset involved implementing various policies and initiatives aimed at reducing the carbon intensity of its economy, increasing energy efficiency, and promoting the use of renewable energy sources such as solar and wind power. Aligned with this new pledge and earlier development that started building up since 2017 [16], we see China strengthening its climate commitments. As global awareness of climate change grew since 2015, China recognised the importance of strengthening its climate commitments.

In autumn 2020, Chinese President Xi Jinping announced during the UN General Assembly that China aimed to peak its carbon emissions before 2030 and achieve carbon neutrality by 2060. This pledge signalled a significant shift in China's climate policy and commitment to long-term sustainability. The opening up a new era for carbon neutrality meant a step forward towards strategies in reducing greenhouse gas emissions and carbon emissions. The plan, as it stands through a long-term development plan, includes targets for reducing carbon emissions across various sectors, transitioning to clean energy sources, investing in green technologies, and enhancing carbon sinks such as forests and wetlands. To support its Carbon Neutrality Plan, China ramped up investments in green technologies and innovation [3, 8, 9]. This includes research and development in areas such as renewable energy, electric vehicles, carbon capture and storage, and smart grid systems [18]. The aim is to develop and deploy cutting-edge technologies to accelerate the transition to a low-carbon economy. Recognising the global nature of climate change, China actively engaged in international cooperation and collaboration to address this issue. It participated in multilateral initiatives such as the Paris Agreement and worked closely with other countries to share knowledge, technology, and best practices for achieving carbon neutrality. Despite its ambitious goals, China faces challenges in transitioning to a carbon–neutral economy, including balancing economic growth with environmental protection, addressing regional disparities, and ensuring social equity [74]. However, achieving carbon neutrality also presents opportunities for innovation, job creation, and sustainable development, which seem to be developing a new business market and mindset in major cities across the country. Hence, the role of core cities has become a bigger highlight in recent years. More importantly, the regional development attributes have become clearer and more integrated in how China could achieve its longer-term plans.

1.2 *Moving Forward: From the Recent Glorious Past of Rapid Development and Urbanisation to Ongoing Pathways of Dual Carbon Neutrality Plan*

We could simply summarise that China's development journey over the past three to four decades has been characterised by rapid urbanisation, economic expansion, and social transformation. From the early reforms of the 1980s to the sustainable development initiatives of the 2010s and the current dual carbon targets, each decade has witnessed significant milestones and policy shifts that have shaped and reshaped China's urban landscape. In a way, we argue about China's development towards making new age of urbanism with a unique mix of development, economic growth, and industrialisation. As China continues to navigate the complexities of urbanisation and development in the twenty-first century, understanding its micro-history provides valuable insights into the challenges and opportunities ahead.

A brief reflection on recent past achievements of China could indicate how China has been marked by remarkable economic growth and urbanisation, lifting millions of people out of poverty and driving the nation to become the world's second-largest economy [18]. This period of rapid development brought unprecedented prosperity and modernisation to many parts of the country, with the construction of vast infrastructure networks, bustling mega-cities, and a burgeoning middle class. China has started to recognise its environmental challenges in a broader sense as well as in detail. Alongside remarkable achievements of its recent past, it came significant environmental challenges, including air and water pollution, habitat destruction, and resource depletion. The essential sacrifices that had to be made to bring up China to where it is now means indicates there are new areas of concern that should be further studied. For instance, the environmental costs of rapid industrialisation and urbanisation became increasingly apparent, threatening public health, ecological balance, and long-term sustainable development [87]. Against this backdrop, China has made a decisive commitment to pursue dual carbon neutrality, aiming to peak carbon emissions before 2030 and achieve carbon neutrality by 2060.

The 2030 targets are happening while China is still urbanising and remains to be a major industrial country. Hence, the complexity is beyond just mere transitions. This signals a fundamental shift in priorities, with the recognition that economic growth must, therefore, be decoupled from carbon emissions and environmental degradation [19]. This brings to light the overall comprehensive action plan, which includes transitioning to renewable energy sources, enhancing energy efficiency, investing in green technology and innovation, and promoting sustainable consumption and production patterns. Crucially, the dual carbon neutrality plan seeks to integrate economic and environmental goals, recognising that sustainable development requires a holistic approach that balances economic prosperity with environmental protection and social equity. This involves reforming economic structures, incentivising green investment and innovation, and aligning policies across sectors to achieve synergistic outcomes.

2 China and the SDGs: A Reflection on China's Sustainable Development and Progress

Since 2015, there is an era for more commitment to sustainability and sustainable development. China's engagement with the Sustainable Development Goals (SDGs) reflects a multifaceted journey towards achieving sustainable development and addressing pressing global challenges, including but to limited to, climate change issues and their impacts on society [48]. As one of the most populous nations and second-largest economy, China plays a crucial role in shaping the trajectory of sustainable development worldwide. A reflection on China's sustainable development and progress within the framework of the SDGs reveals both significant achievements and ongoing challenges [85]. It also means that China has quickly become a role model at the global level.

China's pursuit of sustainable development is deeply rooted in its domestic policies and long-term development strategies. Over the past decades, China has made remarkable strides in poverty alleviation, education, healthcare, and infrastructure development, lifting hundreds of millions of people out of poverty and improving living standards across the country. Yet, there may seem areas that still need further attention to ensure shifting towards people-centered development and create the pathways of achieving the ideals of harmonious development. The so-far achievements align closely with several SDGs, including Goal 1 (No Poverty), Goal 3 (Good Health and Well-being), Goal 4 (Quality Education), and Goal 9 (Industry, Innovation, and Infrastructure).

Moreover, China has demonstrated a strong commitment to environmental protection and climate action, recognising the importance of addressing environmental challenges for sustainable development. Despite many criticism China receives externally, these recent few years created new opportunities that were not there before. We have to bear in mind that China's economy is very much tangled with global economy, and its large-scale production and consumption trends seem inevitable at the current age. Otherwise, as we always say, Santa Claus cannot fulfil a European kid's dream become a reality if China stops producing what it produces now. In a way, we can see that the Chinese government has implemented ambitious policies to tackle air and water pollution, promote renewable energy, and enhance ecological conservation. These efforts contribute to advancing SDGs such as Goal 6 (Clean Water and Sanitation), Goal 7 (Affordable and Clean Energy), Goal 11 (Sustainable Cities and Communities), Goal 13 (Climate Action), and Goal 15 (Life on Land).

China's role in the global economy and its increasing influence on international affairs also have implications for achieving the SDGs beyond its borders. Through initiatives like the Belt and Road Initiative (BRI), China seeks to promote infrastructure development, trade connectivity, and sustainable investment in partner countries [44, 84], contributing to SDGs related to poverty reduction, infrastructure development, and economic growth. However, China's sustainable development journey is not without challenges,and this complexity will be highlighted throughout the book.

2.1 From Slogans to Reality: Foreseeing a Sleeping Dragon

In an event we attended in 2015, we referred to China as a sleeping dragon, particularly in its transitions in sustainable development. Undoubtedly, China's engagement with the SDGs reflects a commitment to sustainable development and progress on multiple fronts. It is not just having a group of slogans but essentialities for the country to become more successful through its almost never-ending development patterns [19]. Yet, it is true that such development pace could not continue forever and there will be an era of stabilising urbanisation and industrial development. And that, we predict to happen after 2030 when China reaches its carbon peak. While significant achievements have been made, addressing remaining challenges requires continued efforts, innovation, and collaboration both domestically and internationally. By aligning its policies and actions with the SDGs, China can further advance its sustainable development agenda and contribute to a more prosperous and equitable world for its citizens and residents.

Aligned with these discussions, we can propose a provocative academic discourse on China's evolving socio-economic and political landscape, transitioning from ideological rhetoric to tangible achievements and potential future trajectories. This discussion delves into the complexities of China's development, exploring the interplay between historical narratives, policy initiatives, and contemporary realities. There are ideological slogans that remain as the backbone of socialist and communism thinking, creating an array of platforms for progress and development—something that may be declining gradually in the so-called democratic states of the West. Throughout its history, China has experienced periods of both dominance and decline, often framed within ideological slogans and narratives propagated by successive regimes. From Mao Zedong's era of revolutionary enthusiasm to Deng Xiaoping's pragmatism and the subsequent "reform and opening-up" policies, China's development has been shaped by ideological rhetoric that sometimes diverged from practical realities [75]. Yet, such mindset has enabled China's transition to become an economic superpower. Some may argue that a foreigner could only naively argue these points, but the reality cannot be simply hidden in China's recent four decades of rapid growth and development. It has been years that China has started challenging established Western hegemony, which indicate the rise of the East among all these established systems of superiority. Deng Xiaoping's market-oriented reforms unleashed China's economic potential, leading to unprecedented growth, industrialisation, and urbanisation, something that was predicted to happen but was not expected to be implemented this fast and at a such scale. The slogan of "socialism with Chinese characteristics" captured this unique blend of state control and market-driven capitalism, which boosted China's rapid economic ascent.

China's rise as a global power carries profound geopolitical implications, sparking debates about the future of international relations and the balance of power. China's Belt and Road Initiative (BRI), military expenditure and modernisation, and assertive foreign policy posture signal its ambition to reshape the global order and assert its influence on the world stage. The juxtaposition of China's domestic challenges with its global ambitions raises questions about the sustainability and implications of its rise. However, we ought to anticipate an era of stabilisation in China's development and growth down the line. As China navigates the transition from slogans to reality, critical questions arise about its future trajectory and the potential for sustainable development, political reform, and social cohesion [18, 63]. Achieving the vision of a rejuvenated "Sleeping Dragon" requires addressing internal contradictions, promoting inclusive growth, fostering innovation, and navigating complex geopolitical dynamics. Ultimately, the trajectory of China's development will shape not only its own destiny but also the contours of the 21st-century global order. Nonetheless, the so-called Chinese Dream (which we cautiously know is only mentioned very briefly here and not to be alienated from the actual discussion) is not far from the reality,it just has to be a different dream that is very much Chinese and for the people of China. Thus, what we see are more than just transitions, and could be discussed or perceived as transformative changes.

3 What Lies Ahead for Cities, Regions, and New Development Zones of China

The future for cities, regions, and new development zones in China holds a mix of opportunities and challenges, shaped by ongoing urbanisation, economic reforms, technological advancements, and sustainability imperatives. We have previously looks at regional development issues very carefully and aim to delve into issues around cities and urban development in recent past and current China. A few directions are highlighted as key pathways to future development, including the development of smart cities and innovation hubs, economic diversification and regional development, sustainable urbanisation and environmental protection that sits well with the overarching high-quality urbanisation plan, integrated transport networks and connectivity, digital governance and e-governance/e-government services, as well as social inclusion and quality of life (Fig. 1). What we can summarise is the future for cities, regions, and new development zones in China is characterized by a dynamic interplay of economic, social, environmental, and technological factors. However, there is no one model, and this must not be the case. We are referring to a country that has extended history, development, and contributions to the globe. We are talking about a populated nation, which is an economic superpower, a role model, and a pioneers in certain development movements or patterns. By embracing innovation, sustainability, and inclusive development principles, China has the potential to build resilient, competitive, and harmonious urban and regional ecosystems that serve as models for global urbanisation in the twenty-first century. There are of course many issues that need to be addressed, and there are still many hills that are yet to be climbed!

4 Aim and Objectives of the Book

Our overarching aim is to provide a mapping study of China's core cities and their changes over three decades of rapid urbanisation, urban growth, and economic development. To portray these very holistically, we explicitly highlight each selected city's development trends, sustainable development plans, and major strategies. We summarise lessons learnt from all 20 case study examples, 10 of which will be covered in this book while the other 10 are covered in the connected forthcoming volume. We hope they can be utilised for other developing and rapidly urbanising nations. We expect this book to be a valuable resource for local governments, authorities, urban planners, urbanists, practitioners, developers, and urban researchers. We trust China

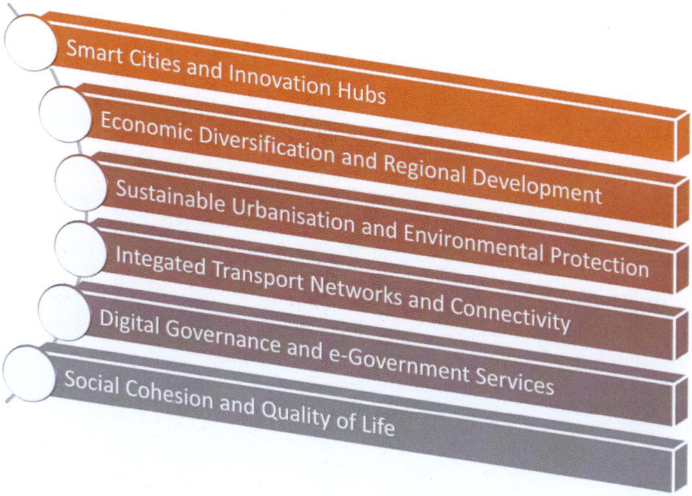

Fig. 1 Summary of foreseeing pathways to future development in China

is no longer a developing country, and much of these recent progressions are owed to its structured urban development strategies, robust governmental structure, and progressive attitude to growth and development. Hence, this topic coverage at the point where China's urbanisation is shifting to high-quality urbanisation is essential and beneficial to multiple stakeholders.

5 Our Comprehensive Selection of China's 20 Core Cities: A Narrative

Rather than just selecting cities randomly or simply sitting down selecting them based on their economic status or population size, we decided to have a procedural selection process. In doing so, we have done a comprehensive ranking based on five different rankings done by different organisations or bodies, including the comprehensive ranking made by (1) China Development Research Foundation (in Chinese: 中国发展研究基金会), from 2017 to 2021, (2) Warton Economic Institute, from 2015 to 2019, (3) Global Netwok Connectivity (GNC) in 2015, which was released in 2016, (4) Global City Index Rank in 2019 based on 2015 data, and (5) Global Cities Outlook Rank in 2019, based on 2015 data. In each category, we identified different main considerable indexes which are highlighted in Table 1.

Tables 2, 3, and 4 summarise ranking from all above-mentioned institutes, organisations, and indexes that are used for our overall ranking shown in Table 5. This process is considered holistic just to ensure our selection is factual and not based on our subjective decision making. This decision making based on multiple rankings enables us to consider a more elaborate approach of multiple years, between

Table 1 Main considerable indexes by different ranking

China Development Research Foundation	Warton Economic Institute	Global Netwok Connectivity (GNC)	Global City Index Rank	Global Cities Outlook Rank
From 2017 to 2021	From 2015 to 2019	2015 (released in 2016)	2015 (released in 2019)	2015 (released in 2019)
• Intellectual capital and innovation • Technical maturity • Major regional cities • Urban resilience • Transportation and urban planning • Sustainable development • Culture and quality of life • Economic clout • Cost • Ease of doing business	• GDP • Saving • Environment • Science and education • Culture • Sanitation	n/a	• Comprehensive	• Leading Industry, Potential

2015 to 2021 (which is the time we started this project). In Table 5, our ranking includes 39 city cases, and we finally select the top 20 cities as core cities in China. This book covers the top 10 cities (Including: Beijing, Shanghai, Shenzhen, Guangzhou, Hangzhou, Chengdu, Nanjing, Wuhan, Tianjin, and Suzhou) which are exemplars of leading urban regions and cities in China. The second volume of these two connected books will cover the remaining 10 cities (Including: Xi'an, Changsha, Zhengzhou, Chongqing, Qingdao, Dalian, Ningbo, Shenyang, Jinan, and Xiamen)—in this suggested order. Together, these two books will cover all our selected 20 Chinese cities.

6 Brief Reflections

At the heart of China's urbanisation progress lies a complex interplay between economic growth, social transformation, and environmental sustainability. The rapid migration of rural populations to urban centers has fuelled economic expansion, driving the country's impressive GDP growth and lifting millions out of

Table 2 Ranking from China Development Research Foundation Data from 2017 to 2021

Rank	2017	2018	2019	2020	2021
1	北京 Beijing	北京 Beijing	北京 Beijing	北京 Beijing	北京 Beijing
2	上海 Shanghai	上海 Shanghai	上海 Shanghai	上海 Shanghai	上海 Shanghai
3	广州 Guangzhou	深圳 Shenzhen	香港 Hongkong	广州 Guangzhou	深圳 Shenzhen
4	深圳 Shenzhen	广州 Guangzhou	广州 Guangzhou	深圳 Shenzhen	广州 Guangzhou
5	杭州 Hangzhou	杭州 Hangzhou	深圳 Shenzhen	成都 Chengdu	杭州 Hangzhou
6	武汉 Wuhan	武汉 Wuhan	杭州 Hangzhou	南京 Nanjing	南京 Nanjing
7	南京 Nanjing	南京 Nanjing	成都 Chengdu	杭州 Hangzhou	成都 Chengdu
8	成都 Chengdu	成都 Chengdu	武汉 Wuhan	武汉 Wuhan	武汉 Wuhan
9	厦门 Xiamen	厦门 Xiamen	厦门 Xiamen	香港 Hongkong	苏州 Suzhou
10	西安 Xian	西安 Xian	南京 Nanjing	郑州 Zhengzhou	香港 Hongkong
11	天津 Tianjin	天津 Tianjin	西安 Xian	长沙 Changsha	长沙 Changsha
12	苏州 Suzhou	长沙 Changsha	长沙 Changsha	苏州 Suzhou	郑州 Zhengzhou
13	珠海 Zhuhai	苏州 Suzhou	苏州 Suzhou	西安 Xian	厦门 Xiamen
14	长沙 Changsha	郑州 Zhengzhou	郑州 Zhengzhou	厦门 Xiamen	西安 Xian
15	郑州 Zhengzhou	珠海 Zhuhai	青岛 Qingdao	重庆 Chongqing	宁波 Ningbo
16	重庆 Chongqing	青岛 Qingdao	济南 Jinan	天津 Tianjin	天津 Tianjin
17	济南 Jinan	济南 Jinan	天津 Tianjin	青岛 Qingdao	青岛 Qingdao
18	昆明 Kunming	昆明 Kunming	澳门 Macao	宁波 Ningbo	重庆 Chongqing
19	青岛 Qingdao	重庆 Chongqing	宁波 Ningbo	昆明 Kunming	无锡 Wuxi

(continued)

Table 2 (continued)

Rank	2017	2018	2019	2020	2021
20	宁波 Ningbo	宁波 Ningbo	昆明 Kunming	济南 Jinan	合肥 Hefei
21	沈阳 Shenyang	无锡 Wuxi	重庆 Chongqing	合肥 Hefei	济南 Jinan
22	大连 Dalian	福州 Fuzhou	福州 Fuzhou	澳门 Macao	昆明 Kunming
23	福州 Fuzhou	贵阳 Guiyang	大连 Dalian	佛山 Foshan	珠海 Zhuhai
24	贵阳 Guiyang	大连 Dalian	珠海 Zhuhai	珠海 Zhuhai	东莞 Dongguan
25	太原 Taiyuan	太原 Taiyuan	合肥 Hefei	无锡 Wuxi	佛山 Foshan
26	无锡 Wuxi	沈阳 Shenyang	佛山 Foshan	太原 Taiyuan	沈阳 Shenyang
27	乌鲁木齐 Urumqi	乌鲁木齐 Urumqi	无锡 Wuxi	贵阳 Guiyang	澳门 Macao
28	石家庄 Shijiazhuang	石家庄 Shijiazhuang	乌鲁木齐Urumqi	沈阳 Shenyang	福州 Fuzhou
29	哈尔滨 Harbin	哈尔滨 Harbin	沈阳 Shenyang	福州 Fuzhou	大连 Dalian
30	兰州 Lanzhou	兰州 Lanzhou	海口 Haikou	大连 Dalian	贵阳 Guiyang

Table 3 Ranking from Warton Economic Institute, from 2015 to 2019

Rank	2017	2018	2019	2020	2021
1	北京 Beijing	北京 Beijing	北京 Beijing	北京 Beijing	北京 Beijing
2	上海 Shanghai	上海 Shanghai	上海 Shanghai	上海 Shanghai	上海 Shanghai
3	广州 Guangzhou	广州 Guangzhou	广州 Guangzhou	广州 Guangzhou	广州 Guangzhou
4	深圳 Shenzhen	深圳 Shenzhen	深圳 Shenzhen	深圳 Shenzhen	深圳 Shenzhen
5	天津 Tianjin	天津 Tianjin	天津 Tianjin	杭州 Hangzhou	杭州 Hangzhou
6	苏州 Suzhou	成都 Chengdu	成都 Chengdu	苏州 Suzhou	苏州 Suzhou
7	武汉 Wuhan	杭州 Hangzhou	杭州 Hangzhou	成都 Chengdu	成都 Chengdu
8	重庆 Chongqing	苏州 Suzhou	苏州 Suzhou	天津 Tianjin	天津 Tianjin
9	杭州 Hangzhou	重庆 Chongqing	重庆 Chongqing	武汉 Wuhan	武汉 Wuhan
10	成都 Chengdu	武汉 Wuhan	武汉 Wuhan	南京 Nanjing	南京 Nanjing
11	南京 Nanjing	沈阳 Shenyang	南京 Nanjing	重庆 Chongqing	重庆 Chongqing
12	沈阳 Shenyang	大连 Dalian	大连 Dalian	长沙 Changsha	长沙 Changsha
13	大连 Dalian	南京 Nanjing	沈阳 Shenyang	郑州 Zhengzhou	郑州 Zhengzhou
14	长沙 Changsha	长沙 Changsha	长沙 Changsha	西安 Xian	青岛 Qingdao
15	无锡 Wuxi	郑州 Zhengzhou	郑州 Zhengzhou	大连 Dalian	无锡 Wuxi
16	西安 Xian	青岛 Qingdao	西安 Xian	无锡 Wuxi	大连 Dalian
17	宁波 Ningbo	西安 Xian	青岛 Qingdao	青岛 Qingdao	西安 Xian
18	郑州 Zhengzhou	无锡 Wuxi	无锡 Wuxi	济南 Jinan	宁波 Ningbo
19	佛山 Foshan	济南 Jinan	济南 Jinan	沈阳 Shenyang	济南 Jinan

(continued)

Table 3 (continued)

Rank	2017	2018	2019	2020	2021
20	济南 Jinan	宁波 Ningbo	宁波 Ningbo	宁波 Ningbo	佛山 Foshan
21	青岛 Qingdao	佛山 Foshan	佛山 Foshan	佛山 Foshan	沈阳 Shenyang
22	长春 Changchun	长春 Changchun	南通 Nantong	南通 Nantong	福州 Fuzhou
23	哈尔滨 Harbin	福州 Fuzhou	哈尔滨 Harbin	长春 Changchun	南通 Nantong
24	烟台 Yantai	烟台 Yantai	东莞 Dongguan	烟台 Yantai	烟台 Yantai
25	石家庄 Shijiazhuang	南通 Nantong	福州 Fuzhou	哈尔滨 Harbin	合肥 Hefei
26	东莞 Dongguan	石家庄 Shijiazhuang	长春 Changchun	石家庄 Shijiazhuang	常州 Changzhou
27	唐山 Tangshan	哈尔滨 Harbin	石家庄 Shijiazhuang	福州 Fuzhou	东莞 Dongguan
28	南通 Nantong	东莞 Dongguan	烟台 Yantai	东莞 Dongguan	昆明 Kunming
29	福州 Fuzhou	唐山 Tangshan	合肥 Hefei	常州 Changzhou	长春 Changchun
30	太原 Taiyuan	太原 Taiyuan	唐山 Tangshan	太原Taiyuan	太原Taiyuan

Table 4 Ranking from Global Netwok Connectivity (GNC) from 2016, Global City Index Rank in 2019, and Global Cities Outlook Rank in 2019

Rank	Global Netwok Connectivity	Global City Index Rank	Global Cities Outlook Rank
1	香港 Hongkong	香港 Hongkong	台北 Taipei (Taiwan)
2	北京 Beijing	北京 Beijing	北京 Beijing
3	上海 Shanghai	上海 Shanghai	深圳 Shenzhen
4	台北 Taipei (Taiwan)	台北 Taipei (Taiwan)	上海 Shanghai
5	广州 Guangzhou	广州 Guangzhou	香港 Hongkong
6	深圳 Shenzhen	深圳 Shenzhen	苏州 Suzhou
7	成都 Chengdu	南京 Nanjing	南京 Nanjing
8	天津 Tianjin	天津 Tianjin	杭州 Hangzhou
9	南京 Nanjing	成都 Chengdu	天津 Tianjin
10	杭州 Hangzhou	杭州 Hangzhou	西安 Xian
11	青岛 Qingdao	苏州 Suzhou	武汉 Wuhan
12	大连 Dalian	武汉 Wuhan	无锡 Wuxi
13	重庆 Chongqing	重庆 Chongqing	广州 Guangzhou
14	厦门 Xiamen	大连 Dalian	东莞 Dongguan
15	台州 Taizhong (Taiwan)	西安 Xian	大连 Dalian
16	武汉 Wuhan	青岛 Qingdao	沈阳 Shenyang
17	苏州 Suzhou	长沙 Changsha	哈尔滨 Harbin
18	长沙 Changsha	哈尔滨 Harbin	宁波 Ningbo
19	西安 Xian	宁波 Ningbo	成都 Chengdu
20	沈阳 Shenyang	沈阳 Shenyang	郑州 Zhengzhou
21	济南 Jinan	郑州 Zhengzhou	佛山 Foshan
22	高雄 Kaohsiung (Taiwan)	无锡 Wuxi	唐山 Tangshan
23	昆明 Kunming	佛山 Foshan	重庆 Chongqing
24	福州 Fuzhou	烟台 Yantai	青岛 Qingdao
25	澳门 Macao	东莞 Dongguan	长沙 Changsha
26	太原 Taiyuan	唐山 Tangshan	烟台 Yantai
27	长春 Changchun	--	--
28	合肥 Hefei	--	--
29	宁波 Ningbo	--	--
30	郑州 Zhengzhou	--	--

poverty. Urbanisation has also facilitated technological advancements, innovation, and improved standards of living for many Chinese citizens. All these together have enabled new opportunities for a rapidly-developing country as well as other nations and regions who are connected with such growth and development. Moreover, fostering sustainable urbanisation in China requires collaboration between government, businesses, academia, and civil society. By fostering partnerships and engaging stakeholders at all levels, China can leverage diverse expertise and resources to address complex sustainability challenges effectively. Ultimately, China's urbanisation progress represents a critical confluence in its development trajectory. By embracing sustainability as a guiding principle, China can build cities that not only thrive economically but also nurture the well-being of its citizens and safeguard the environment for future generations. The path to sustainable urbanisation may be challenging, but it is essential for securing a prosperous and harmonious future for China and the world.

This book delves into 10 selected core cities in China (out of the final list of 20 cities), all covered in the remaining chapters of the book. The idea is to explore

Table 5 Our final comprehensive ranking based on all available data from 2015 to 2021

References Total

Volume 1 (Covered in this book): ranks 1–10

Volume 2 (Covered in next book): ranks 11–20

Rank	City	Score
1	北京 Beijing	387
2	上海 Shanghai	373
3	深圳 Shenzhen	349
4	广州 Guangzhou	347
5	杭州 Hangzhou	316
6	成都 Chengdu	297
7	南京 Nanjing	289
8	武汉 Wuhan	281
9	天津 Tianjin	276
10	苏州 Suzhou	274
11	西安 Xian	221
12	长沙 Changsha	210
13	郑州 Zhengzhou	194
14	重庆 Chongqing	194
15	青岛 Qingdao	183
16	大连 Dalian	166
17	宁波 Ningbo	149
18	沈阳 Shenyang	141
19	济南 Jinan	122
20	厦门 Xiamen	118
21	佛山 Foshan	79
22	昆明 Kunming	69
23	福州 Fuzhou	67
24	哈尔滨 Harbin	57
25	珠海 Zhuhai	56
26	东莞 Dongguan	52
27	合肥 Hefei	42
28	烟台 Yantai	38
29	无锡 Wuxi	37
30	长春 Changchun	37
31	南通 Nantong	36
32	太原 Taiyuan	27
33	贵阳 Guiyang	26
34	石家庄 Shijiazhuang	26
35	唐山 Tangshan	20
36	乌鲁木齐 Urumqi	11
37	常州 Changzhou	7
38	兰州 Lanzhou	2
39	海口 Haikou	1

these cases in further detail, evaluate their micro-history of development, map their development patterns, and see what measures and strategies are in place to achieve sustainability and sustainable development. In doing so, our aim in this round is to focus on city-level where we believe difficulties of implementing low carbon, smart, green, eco, and other projects have become a growing challenge. While demand in such directions has grown extensively, we note challenges are equally piling up, indicating that our fast actions may not be necessarily right solutions for our long-term problems. In some ways, China is partially in its learning curve and is still testing a few pilot projects. But we also see scaling up projects or major initiatives that are transformative and impactful down the line. The book is not aiming to provide a biased narrative of these 10 case study examples, but explore their progress, achievements, challenges, and pathways; those that we believe are essential for China and other countries, too. Cities of smaller scale will look up to such role models, while the world is watching what China will do or do not do in these coming years and decades. China's path to become a global actor has set a unique opportunity to show a good mix of desire, speed, size, and efforts all happening under these unprecedented processes of urban change across the nation. We agree that China is no longer the so-called developing nation but is an emerging power that for long has been waiting to get this far.

Much effort is put in place to make this book a reality. We went through ups and downs of 2021 to 2024 to make sure all our work is developed accurately, cohesively, and rationally. Disruptions occurred and we acknowledge them very clearly, but our vision of creating a holistic book for readers remained solid in these few no-to-easy years. We had several rounds of excellent discussions, data collection, and case study reviews, visiting many stakeholders in most of our selected cities, and then ensuring all are documented based on facts and a right narrative that indicate China's progress is owed to both rural and urban populations—and not cities/city environments or urban areas alone. The two have been equally important, and the economic expansion and growth of cities could have never happened if rural production did not support the country's overarching long-term vision and plans. As we focus on urban areas and hubs, we try to avoid focusing on mere economic development patterns. The complexity of sustainable development directions in Chinese cities indicates the fact that there is more than just one sustainable development pathway. If to see sustainability as one vision, then we cannot see much of diversity and uniqueness in cities and their development models or patterns. Thus, we hope this book is an important learning resource for those who seek to understand urban change and its importance in sustainable development. If to neglect these matters and only focus on generic urban design and planning solutions, then we better give all our ongoing and demanding tasks to an AI-assisted platform. It could perhaps generate a more creative urban development model that what we could come up with if we only have one solution in mind. However, this is not what we expect to happen, and this is not what makes a people-centric urban development we seek for in the future. A direction towards high-quality urbanisation, as China's plan indicates now, means getting away from egocentric models, having a better knowledge of our nature and the environments, consider people and our eco-systems at the core of our development patterns, and ensure we have an all-inclusive mindset to sustainable development that could actually sustain and become tangible and meaningful for all.

References

1. Agarwal J, Wu T (2004) China's entry to WTO: global marketing issues, impact, and implications for China. Int Mark Rev 21(3):279–300
2. Ahlers A (2014) Rural policy implementation in contemporary China: New socialist countryside. Routledge, New York
3. Ahmed K (2023) Perspective on China's commitment to carbon neutrality under the innovation-energy-emissions nexus. J Clean Prod 390:136202
4. Amsden AH (2009) Escape from empire: the developing world's journey through heaven and hell. MIT Press, Cambridge
5. Barton D, Chen Y, Jin A (2013) Mapping China's Middle Class. Mckinsey Q 3(2013):54–60
6. Bhardwaj RD (1992) China's economic reform: the role and significance of SEZs. Indian J Polit Sci 53(3):332–373
7. Brandt L, Ma D, Rawski TG (2014) From divergence to convergence: reevaluating the history behind China's economic boom. Am Econo J J. Econ Lit 52(1):45–123
8. Busch C, Jun M, Harvey H, Min H (2021) China's carbon neutral opportunity. Energy innovation: policy and technology, LLC.: San Francisco, CA, USA
9. Can W, Ruoshui S, Jiutian Z (2021) Supportive technologies and roadmap for China's carbon neutrality. China Economist 16(5):32–70
10. Cao GY, Chen G, Pang LH, Zheng XY, Nilsson S (2012) Urban growth in China: past, prospect, and its impacts. Popul Environ 33:137–160
11. Chan KW (1992) Economic growth strategy and urbanization policies in China, 1949–1982. Int J Urban Reg Res 16(2):275–305
12. Chen C, LeGates R, Zhao M, Fang C (2018) The changing rural-urban divide in China's megacities. Cities 81:81–90
13. Chen W, Cheshmehzangi A, Mangi E, Heath T (2022) Implementations of China's new-type urbanisation: A comparative analysis between targets and practices of key elements' policies. Sustainability 14(10):6341
14. Chen W, Zhang S, Kong D, Zou T, Zhang Y, Cheshmehzangi A (2023) Spatio-temporal evolution and influencing factors of China's ICT service industry. Sci Rep 13(1):9703
15. Cheshmehzangi A (2016) China's New-Type Urbanisation Plan (NUP) and the foreseeing challenges for decarbonization of cities: a review. Energy Procedia 104:146–152
16. Cheshmehzangi A (2020) Towards urban energy management: a brief review of China's recent transition in urban energy demand. Low Carbon Economy 11(01):1
17. Cheshmehzangi A, Chen H (2021) China's sustainability transitions: low carbon and climate-resilient plan for carbon neutral 2060. Springer Nature, Singapore
18. Cheshmehzangi A, Dawodu A, Sharifi A (2021) Sustainable Urbanism in China. Routledge, New York and London
19. Cheshmehzangi A, Tang T (2022) China's city cluster development in the race to carbon neutrality. Springer, Singapore
20. Cheshmehzangi A, Xie L, Tan-Mullins M (2018) The role of international actors in low-carbon transitions of Shenzhen's International Low Carbon City in China. Cities 74:64–74
21. Cheshmehzangi A, Zou T (2022) Revealing the new type of urban displaced people through the lens of the COVID-19 pandemic: the neglected vulnerable groups during lockdown. Bhumi, Plan Res J 9(1)
22. Chiu RL (2012) Urban sustainability and the urban forms of China's leading mega cities: Beijing, Shanghai and Guangzhou. Urban Policy Res 30(4):359–383

23. Croll E (2006) China's new consumers: social development and domestic demand. Routledge, New York and London
24. Deng W, Cheshmehzangi A (2018) Eco-development in China: cities, communities and buildings. Palgrave Macmillan, Singapore
25. Dmitrienko A, Lipovenko M, Gostilovich A, Kolosov A, Ming L (2023) Development strategies of high-tech companies in China: Huawei and Tencent. Nexo Revista Científica 36(02):176–187
26. Economy E, Levi M (2014) By all means necessary: How China's resource quest is changing the world. Oxford University Press, Oxford
27. Eichengreen B, Tong H (2006) How China is reorganizing the world economy. Asian Economic Policy Review 1(1):73–97
28. Feng W (2018) Rural–urban migration, urbanization, and megacities in China. In: The rise of megacities: challenges, opportunities and unique characteristics, p 175
29. Gálvez L, Cheshmehzangi A (2015) China's urban housing: the review of three housing typologies and patterns. In: Proceedings of the 6th conference on urban space and social life: theory and practice, Macau, China, pp 6–11, June 2015
30. Gerth K (2010) As China goes, so goes the world: how Chinese consumers are transforming everything. Hill and Wang, New York
31. Golley J, Song L (2010) Chinese economic reform and development: achievements, emerging challenges and unfinished tasks. China: The next twenty years of reform and development, pp 1–18
32. Goodman DS (2004) The campaign to "Open up the West": national, provincial-level and local perspectives. The China Quarterly 178:317–334
33. Gonzalez-Vicente R (2011) The internationalization of the Chinese state. Polit Geogr 30(7):402–411
34. Gu C, Kesteloot C, Cook IG (2015) Theorising Chinese urbanisation: A multi-layered perspective. Urban Studies 52(14):2564–2580
35. Hansen A (2020) Consumer socialism: Consumption, development and the new middle classes in China and Vietnam. The socialist market economy in Asia: Development in China, Vietnam and Laos, pp 221–243
36. Harrison J, Gu H (2021) Planning megaregional futures: Spatial imaginaries and megaregion formation in China. Planning Regional Futures. Routledge, New York, pp 148–171
37. He C, Chen T, Mao X, Zhou Y (2016) Economic transition, urbanization and population redistribution in China. Habitat Int 51:39–47
38. Heilmann S (2008) Policy experimentation in China's economic rise. Stud Comp Int Dev 43(1):1–26
39. Holbig H (2004) The emergence of the campaign to open up the West: Ideological formation, central decision-making and the role of the provinces. China Q 178:335–357
40. Jiang GH (2023) Pursuing Modernisation in China. The imperial mode of China: an analytical reconstruction of Chinese economic history. Springer International Publishing, Cham, pp 255–304
41. Kamal-Chaoui L, Leeman E, Rufei Z (2009) Urban trends and policy in China. OECD Regional Development Working paper, at OECD iLibrary. Available from: https://www.oecd-ilibrary.org/content/paper/225205036417
42. Kennedy S (2010) The myth of the Beijing Consensus. J Contemp China 19(65):461–477
43. Kennedy S (2020) China's uneven high-tech drive. Center for Strategic and International Studies (CSIS). Available from: http://csis-website-prod.s3.amazonaws.com/s3fs-public/publication/200302_Kennedy_ChinaUnevenDrive_v3.pdf
44. Khan MK, Sandano IA, Pratt CB, Farid T (2018) China's Belt and Road Initiative: a global model for an evolving approach to sustainable regional development. Sustainability 10(11):4234
45. Knoth C (2000) Special Economic Zones and economic transformation. The case of the People's Republic of China. Dissertation document. Available from: https://kops.uni-konstanz.de/handle/123456789/12261

46. Kotler P, Kotler M (2014) Winning global markets: how businesses invest and prosper in the world's high-growth cities. Wiley, New York
47. Kroeber AR (2020) China's economy: what everyone needs to know. Oxford University Press, Oxford
48. Kuhn BM (2018) China's commitment to the sustainable development goals: an analysis of push and pull factors and implementation challenges. Chin. Polit. Sci. Rev. 3(4):359–388
49. Kynge J (2010) China shakes the world: The rise of a hungry nation. Pheonix, Hachette, UK.
50. Li C (1997) Rediscovering China: Dynamics and dilemmas of reform. Rowman & Littlefield, Lanham
51. Li R, Huang X, Liu Y (2023) Spatio-temporal differentiation and influencing factors of China's urbanization from 2010 to 2020. Acta Geogr Sin 78:777–791
52. Liang G (2004) New competition; foreign direct investment and industrial development in China (No. ERIM PhD Series Research in Management; EPS-2004-047-ORG)
53. Liu Y, Yau Y (2020) Urban entrepreneurialism vs market society: the geography of China's neoliberal urbanism. Int J Urban Reg Res 44(2):266–288
54. Logan J (ed) (2011) Urban China in transition. Wiley, New York
55. Lu R (2008) Chinese special economic zones as clusters: a case study of Shenzhen's modern service clusters. Master's thesis, Høgskolen i Bodø
56. Meyer-Clement E (2016) The great urban leap? On the local political economy of rural urbanisation in China. J Curr Chin Aff 45(1):109–139
57. Miller T (2012) China's urban billion: the story behind the biggest migration in human history. Bloomsbury Publishing, London
58. Minzner C (2018) End of an era: How China's authoritarian revival is undermining its rise. Oxford University Press, Oxford
59. Nguyen BG (2022) Agriculture in communist China: progress without structural reforms. China Report 58(3):300–316
60. Pétursson E (2012) China's urban development in the post-reform era. The Chinese city and the emerging mega city in wake of the open door policy. Doctoral dissertation. Available from: https://skemman.is/handle/1946/12864
61. Pow CP (2018) Building a harmonious society through greening: Ecological civilization and aesthetic governmentality in China. Ann Am Assoc Geogr 108(3):864–883
62. Ren Y (2021) Rural China in the digital era: evolution, opportunities and challenges. In: International symposium on advancement of construction management and real estate. Springer Nature Singapore, Singapore, pp 1347–1358, November.
63. Saich T (2017) Governance and politics of China. Bloomsbury Publishing, London
64. Schiere R (2009) China's development challenges: economic vulnerability and public sector reform. Routledge, New York
65. Shambaugh D (2016) China's future. Wiley, New York
66. Shih V (2004) Development, the second time around: the political logic of developing western China. J. East Asian Stud 4(3):427–451
67. Shi Q, Cao G (2020) Urban spillover or rural industrialisation: which drives the growth of Beijing Metropolitan Area. Cities 105:102354
68. Song E (1996) Shanghai: transformation and modernization under China's open policy. Chinese University Press, Beijing
69. Teo HC, Lechner AM, Walton GW, Chan FKS, Cheshmehzangi A, Tan-Mullins M, Chan HK, Sternberg T, Campos-Arceiz A (2019) Environmental impacts of infrastructure development under the belt and road initiative. Environments 6(6):72
70. Tilt CA (2016) Corporate social responsibility research: the importance of context. Int J Corp Soc Responsib 1:1–9
71. Tuan C, Ng LF (2004) Manufacturing agglomeration as incentives to Asian FDI in China after WTO. J Asian Econ 15(4):673–693
72. Veeck G (2007) China's geography: globalization and the dynamics of political, economic, and social change. Rowman & Littlefield, Lanham

73. Wang HH (2010) The Chinese dream: the rise of the world's largest middle class and what it means to you. Best Seller Press, Augustine
74. Wang S, Chen F, Wang Y (2023) Low-carbon technological innovation of coal production and utilization impetus carbon neutrality in China. Environ Sci Pollut Res 30(33):80916–80930
75. Wei CN (2019) FROM MAO TO DENG TO XI. Int J World Peace 36(2):31–58
76. Wei H, Li L, Nian M (2021) China's urbanization strategy and policy during the 14th five-year plan period. Chin J Urban Environ Stud 9(01):2150002
77. Wei H, Wei, Li (2019) Urbanization in China. Springer, Singapore.
78. Wong YC (2005) From Deng Xiaoping to Jiang Zemin: two decades of political reform in the People's Republic of China. University Press of America, Lanham
79. Wu F (2020) Adding new narratives to the urban imagination: An introduction to 'New directions of urban studies in China.' Urban Stud 57(3):459–472
80. Xie L, Cheshmehzangi A, Tan-Mullins M, Flynn A, Heath T (2020) Urban entrepreneurialism and sustainable development: a comparative analysis of Chinese eco-developments. J Urban Technol 27(1):3–26
81. Xie L, Tan-Mullins M, Cheshmehzangi A (2019) Political ecology of Chinese smart eco-cities. Remaking sustainable urbanism: Space, scale and governance in the new urban era, 57–78
82. Xu W, Tan KC (2002) Impact of reform and economic restructuring on rural systems in China: a case study of Yuhang, Zhejiang. J Rural Stud 18(1):65–81
83. Ye Q, Qijiao S, Xiaofan Z, Shiyong Q, Lindsay T (2020) China's new urbanisation opportunity: a vision for the 14th Five-Year Plan. Coalition for Urban Transitions, London, UK
84. Yin W (2019) Integrating Sustainable Development Goals into the Belt and Road Initiative: would it be a new model for green and sustainable investment? Sustainability 11(24):6991
85. Yu S, Sial MS, Tran DK, Badulescu A, Thu PA, Sehleanu M (2020) Adoption and implementation of sustainable development goals (SDGs) in China—Agenda 2030. Sustainability 12(15):6288
86. Yu Z, Jiang S, Cheshmehzangi A, Liu Y, Deng X (2023) Agricultural restructuring for reducing carbon emissions from residents' dietary consumption in China. J Clean Prod 387:135948
87. Yuan J, Lu Y, Ferrier RC, Liu Z, Su H, Meng J, Song S, Jenkins A (2018) Urbanization, rural development and environmental health in China. Environ Dev 28:101–110
88. Zhang L, LeGates R, Zhao M (2016) Understanding China's urbanization: the great demographic, spatial, economic, and social transformation. Edward Elgar Publishing
89. Zhang S (2016) Industrialising China, escaping labour: economic development and the agency of migrant labour in Guangdong, Zhejiang and Jiangsu province. Doctoral dissertation, SOAS University of London
90. Zhang S, Zhao J, Jiang Y, Cheshmehzangi A, Zhou W (2023) Assessing the rural–urban transition of China during 1980–2020 from a coordination perspective. Land 12(6):1175
91. Zhang WW (2000) Transforming China. Palgrave Macmillan UK, London
92. Zhao SX, Guo NS, Li CLK, Smith C (2017) Megacities, the world's largest cities unleashed: major trends and dynamics in contemporary global urban development. World Dev 98:257–289
93. Zhou Y (2008) The inside story of China's high-tech industry: Making Silicon Valley in Beijing. Rowman & Littlefield, Lanham
94. Zou L, Shen JH, Zhang J, Lee CC (2022) What is the rationale behind China's infrastructure investment under the Belt and Road Initiative. J Econ Surv 36(3):605–633

Chapter 2
Beijing: Balancing Tradition and Modernisation in a Metropolis with Multiple Agenda on Sustainability and Innovation

1 Introduction to Beijing: Beyond Just a Historical Status

The People's Republic of China's capital, Beijing, is a dynamic city with a rich cultural legacy. Beijing, the nation's political, cultural, and educational center, is well situated in the country's north. Beijing is a major transportation hub that links the various regions of the nation [5]. Its advantageous location strengthens its position as a key participant in China's economic growth and cross-cultural interchange. In the past few decades, Beijing has undergone a remarkable transformation, evolving into a global powerhouse of economic development, innovation, and urbanisation [8]. Partly due to its political status and historical values, it has grown to be an influential city in the northern regions of China. The city's development strategies have been carefully crafted to propel it into the ranks of the world's leading metropolises while preserving its rich cultural heritage. One of the key strategies driving Beijing's development in recent decades has been urban planning and infrastructure investment. The city has embarked on ambitious projects to modernise its transportation networks, expand its urban area, and create more sustainable living environments. This includes the construction of extensive subway systems, modern highways, and the iconic Beijing Daxing International Airport, which has become one of the busiest airports in the world.

In addition to physical infrastructure, Beijing has also focused on fostering a conducive environment for innovation and technology [7]. The city has invested heavily in research and development, attracting top talent and fostering collaboration between academia, government, and industry. This has led to the emergence of thriving tech hubs such as Zhongguancun [38], often referred to as China's Silicon Valley [48], where cutting-edge technologies are developed and commercialised.

Beijing's historical narrative represents its today's image of a cultural and political city. Throughout its three millennia of history, Beijing has seen both revolutionary upheaval and imperial majesty. Although its origins are in the Warring States era, it became well-known as the capital throughout the Ming, Qing, and Yuan dynasties.

A. Cheshmehzangi and T. Tang, *30 Years of Urban Change in China's 10 Core Cities*, Urban Sustainability, https://doi.org/10.1007/978-981-97-8846-0_2

The city's historical significance is reflected in the magnificent imperial palace known as the Forbidden City. In the twentieth century, Beijing witnessed pivotal events, including the establishment of the People's Republic of China in 1949.

1.1 General Information: From Cultural Legacy to Economic Diversification

Displaying a population of over 21 million, Beijing is a cultural center in addition to a political powerhouse. The city's cultural legacy is exhibited by its architectural wonders, which include the Great Wall, the Temple of Heaven, and the Summer Palace [24]. Beijing's skyline, which includes modern buildings like the Bird's Nest and Water Cube from the 2008 Olympics, perfectly blends heritage and modernity.

Furthermore, Beijing has pursued policies to promote economic diversification and sustainable growth. The city has encouraged the development of strategic industries such as finance, information technology, and advanced manufacturing, while also supporting small and medium-sized enterprises (SMEs) [43] and startups [41]. Initiatives like the Beijing Economic-Technological Development Area (BDA) have provided a conducive environment for businesses to thrive, driving job creation and economic prosperity [49]. Another crucial aspect of Beijing's development strategy has been its emphasis on environmental sustainability and ecological conservation [11]. Recognising the importance of addressing environmental challenges such as air pollution and resource depletion, the city has implemented stringent regulations, invested in clean energy infrastructure, and promoted green initiatives [9]. Efforts such as the conversion of coal-fired heating systems to cleaner alternatives and the expansion of green spaces and parks have contributed to improving the quality of life for residents.

Moreover, Beijing has sought to enhance its global connectivity and influence through initiatives like the Belt and Road Initiative (BRI) and the hosting of international events such as the 2008 Summer Olympics and the Belt and Road Forum for International Cooperation [25]. These efforts have not only bolstered Beijing's reputation as a global city but have also facilitated economic cooperation and cultural exchange with other nations. One of the most significant endeavors in this regard is China's Belt and Road Initiative (BRI), a vast infrastructure and economic development project spanning Asia, Africa, and Europe. As the capital of China, Beijing plays a central role in driving and coordinating BRI activities [35], [11]. The city serves as a focal point for planning, financing, and implementing BRI projects, leveraging its institutional resources, financial institutions, and diplomatic networks to facilitate cross-border cooperation [10]. By actively participating in the BRI, Beijing aims to strengthen economic ties with partner countries, expand market access for Chinese businesses [33], and promote regional stability and development [14].

Perhaps, more importantly, Beijing serves as the political hub where important choices affecting China's future are decided. Beyond politics, the city is important for international relations, technology, and education. It hosts prestigious institutions like Peking University and Tsinghua University, contributing to its global academic standing. Furthermore, Beijing has actively engaged in bilateral and multilateral partnerships to enhance its global connectivity and influence; i.e., in a sort of institutional-balancing strategy [46]. The city has established sister-city relationships with major capitals and metropolises worldwide, facilitating cooperation in areas such as trade, education, tourism, and technology transfer. By forging strategic alliances and collaborative networks, Beijing aims to promote mutual benefit, shared prosperity, and sustainable development across borders.

2 Beijing and Its Physical and Main Development Over the Past 30 Years

Beijing has changed dramatically in the last thirty years, becoming a contemporary global metropolis. Its infrastructure and skyline have changed due to rapid urbanisation and economic growth. Modern transportation networks, high-rise structures, and new districts are all signs of the city's growth. Beijing's global prominence is symbolized by iconic landmarks like the Water Cube and Bird's Nest, which were built because of initiatives like the 2008 Olympics. Figures 1, 2, 3 and 4 show Beijing's development over these thirty years of change.

The development of transportation infrastructure has been a key focus area for Beijing's urban planners. The city has invested heavily in expanding its subway network, which has become one of the largest and busiest in the world, providing convenient and efficient transportation for millions of commuters every day. Additionally, Beijing has developed an extensive network of highways and expressways, easing traffic congestion and improving connectivity within the city and with neighboring regions [14]. Aligned with its extensive infrastructure development, Beijing has also experienced rapid economic growth and industrialisation. The city has become a major hub for finance, technology, and innovation, attracting domestic and foreign investment and fostering the growth of vibrant business ecosystems [28]. Efforts to transition to clean energy sources, enhance green spaces, and promote eco-friendly transportation have been central to Beijing's agenda for sustainable development.

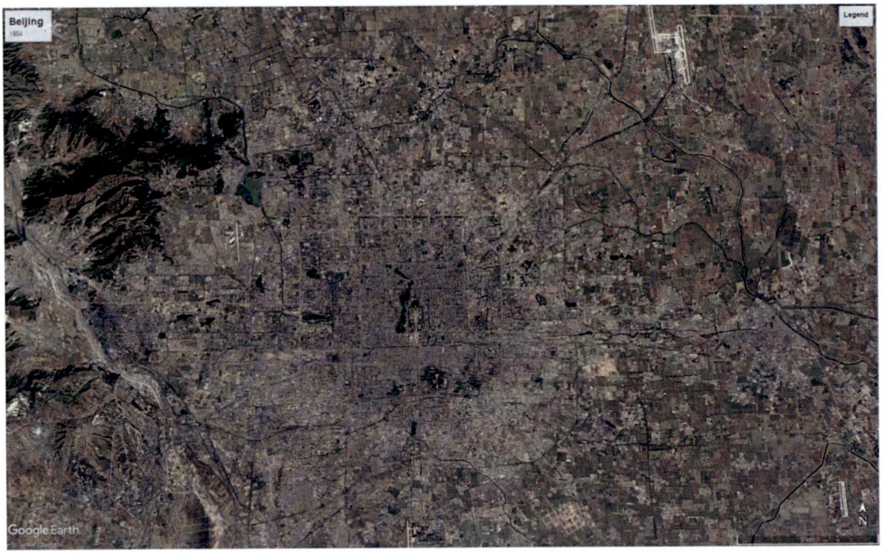

Fig. 1 Beijing Development in 1994 (map is extracted from open and accessible maps on Google Earth, generated in April 2024)

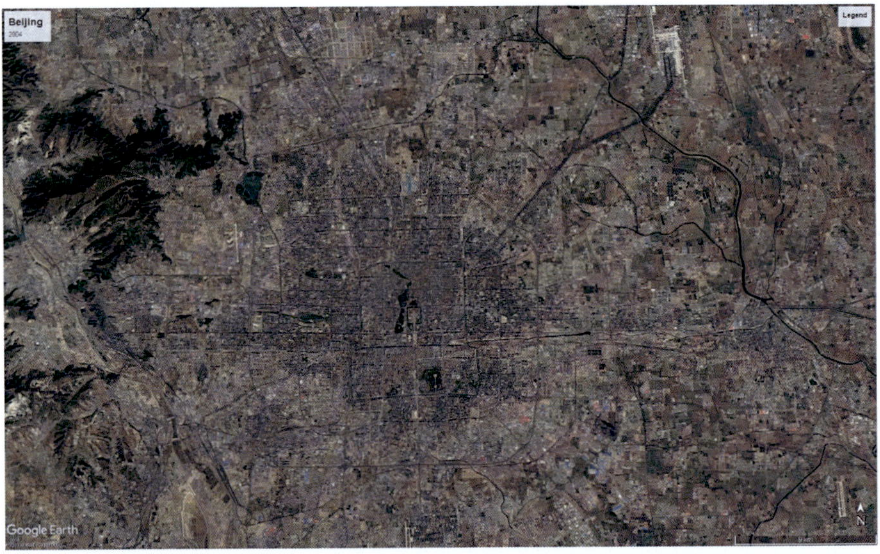

Fig. 2 Beijing Development in 2004 (map is extracted from open and accessible maps on Google Earth, generated in April 2024)

Fig. 3 Beijing Development in 2014 (map is extracted from open and accessible maps on Google Earth, generated in April 2024)

Fig. 4 Beijing Development in 2024 (map is extracted from open and accessible maps on Google Earth, generated in April 2024)

2.1 *Distribution of Residential Neighbourhoods Through Clear Characterisation*

The distribution of residential neighborhoods in Beijing reflects a combination of geographical factors, historical development patterns, and urban planning initiatives. The prevalence of residential areas in the north compared to the south can be attributed to the city's geographic layout. With mountainous terrain in the west, development was historically limited, leading to lower population density in that region. In contrast, the northern areas experienced early urbanization and development, resulting in a more continuous distribution of residential neighborhoods [50]. East is relying on the CBD, but also open plains, extending the main roads and subway lines, residential neighborhoods dense, has been stretched to Tongzhou. This expansion is driven by factors such as land availability, infrastructure investment, and government policies aimed at decentralizing the city center and alleviating congestion.

Similarly, the east–west distribution of residential areas is influenced by Beijing's geographical features and urbanization trends. The eastern part of the city, particularly around the Central Business District (CBD), has seen rapid development due to its flat terrain and proximity to economic centers [26, 30]. As a result, residential neighbourhoods have increased in the east, supported by the extension of main roads and subway lines to accommodate the growing population. By understanding these factors, city planners can better anticipate future growth trends and develop sustainable strategies for accommodating Beijing's evolving population needs.

2.2 *Metro Planning Aligned with Beijing's Population Distribution Strategies*

The metro planning in Beijing intricately aligns with the city's population distribution dynamics, as evidenced by the integration of residential areas with subway networks on maps. Notably, Beijing's subway expansion strategy mirrors the patterns of population growth and urban development [1]. Following the successful bid for the Olympics (Fig. 5), a surge in development activity was observed in the northern region of Beijing. Consequently, new subway lines were strategically extended to serve this burgeoning area, enhancing connectivity and accessibility. Subsequently, in 2009, Beijing unveiled plans aimed at promoting the development of the southern part of the city, signaling a deliberate effort to balance growth across different geographic sectors [16]. This initiative was accompanied by a commitment to expand subway infrastructure in the southern districts, reflecting a proactive approach to accommodate anticipated population increases and support sustainable urbanisation.

Moreover, with the relocation of the city government to the eastern sector, there is a prevailing expectation that transportation infrastructure development will gain momentum in the east. This anticipation underscores the integral role of transportation planning in shaping urban expansion and connectivity [47]. As Beijing continues

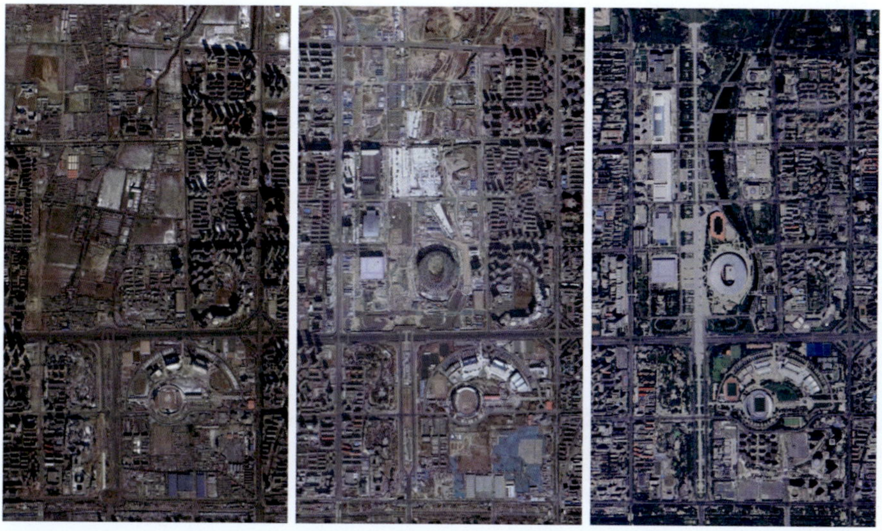

Fig. 5 Beijing Olympic Park and surrounding area from early construction stage in the 2000s till 2024 (maps are extracted from open and accessible maps on Google Earth, generated in April 2024)

to evolve and grow, the strategic expansion of the subway network will play a pivotal role in facilitating efficient mobility, reducing congestion, and promoting balanced regional development.

2.3 Three Business Districts to Form a Polycentric Development Approach

The heat distribution map of Beijing's office area stock shows that Beijing's three major business districts are CBD, Financial Street and Zhongguancun. Others such as Fengtai Science and Technology Park, Wangjing, Ya'ao, and Liangguang Street also have a certain amount, but have not formed a large-scale business district. The three major business districts are home to the largest number of financial, information, and high-tech headquarters in China [44], which not only gradually form the pattern of Beijing's polycentric or multi-center development, but also provide many high-end employment opportunities [23]. These business concentrations not only delineate the polycentric or multi-center development pattern of Beijing but also furnishes a wealth of high-end employment opportunities.

The CBD serves as the focal area for diverse corporate activity, housing multinational corporations, financial institutions, and government agencies [20]. Financial Street, renowned as the financial heart of Beijing, accommodates major banks, investment firms, and regulatory bodies. Zhongguancun, often dubbed China's Silicon Valley, is a breeding ground for technological innovation, boasting numerous research institutions, startups, and tech giants [27]. This clustering of industry leaders fosters collaboration, innovation, and knowledge exchange, boosting Beijing's status as a global economic hub. Moreover, the presence of these business districts catalyses infrastructural development, urban revitalisation, and cultural enrichment, thereby contributing to Beijing's vibrancy and dynamism on the world stage.

3 Beijing's Development Strategies: From Cultural and Educational Districts to the Nation's Leading Cultural and Educational Resources

From the recent financial reports [31], it can be seen that the cultural and educational districts are mainly concentrated in Zhongguancun, and Beijing also has the largest number of universities and the highest quality universities in the country. In particular there is rapid growth in certain industries and manufacturing talents and opportunities. If the three major business districts are the dream places for young people pursuing their careers, then these colleges and universities are the dream places for young people all over the country. This is the common slogan you see in many places, indicating that Beijing is rapidly becoming a major magnet point for younger generations. Nevertheless, the growth is partially different to what has been anticipated for the city's population growth and distribution plans.

Beijing stands as a beacon of cultural and educational richness, displaying a myriad of districts that serve as the nation's leading cultural and educational resources. From ancient landmarks to modern institutions, the city's diverse districts are hubs of creativity, learning, and intellectual exchange. For instance, at the heart of Beijing lies the historic center, where iconic landmarks such as the Forbidden City, Temple of Heaven, and Tiananmen Square bear witness to millennia of Chinese history and culture. These sites not only serve as tourist attractions but also as invaluable educational resources, offering insights into China's rich cultural heritage and historical legacy. Beyond the ancient heart of the city, Beijing is home to vibrant cultural districts such as 798 Art Zone and Dashanzi Art District, where contemporary art galleries, studios, and exhibitions showcase the latest trends in art and design [40]. These districts serve as incubators for artistic expression and cultural innovation, attracting artists, scholars, and enthusiasts from around the world. More importantly, by preserving its cultural legacy, fostering artistic creativity, and nurturing academic talent, Beijing continues to play a central role in shaping the nation's cultural identity and intellectual development.

Finally, we can argue that Beijing's development strategies indicate a set of strategic positioning of the city as a national political center, cultural center, center of international communication, and center of scientific and technological innovation [4]. Firstly, as the national political center, Beijing's policies and initiatives are geared towards upholding its status as the capital of China, where the country's central government institutions, including the National People's Congress and the

State Council, are headquartered [14]. Secondly, Beijing aims to solidify its role as a cultural center by preserving and promoting its rich cultural heritage, manifested in its historic landmarks, museums, theaters, and traditional arts. The city's cultural initiatives not only celebrate its past but also foster cultural exchange and creativity, enhancing its cultural influence both domestically and internationally. Lastly, Beijing is committed to becoming a leading center of scientific and technological innovation, as evidenced by its investments in research and development [19], the establishment of technology parks such as Zhongguancun, and initiatives to attract top talent and foster entrepreneurship. By leveraging its intellectual capital and technological prowess, Beijing aims to drive economic growth, address societal challenges, and contribute to global innovation networks.

3.1 Beijing and Its Overarching 2020 Development Goals

Substantial achievements were seen under the establishment of earlier (but recent) goals to have an internationally acclaimed, harmonious, and liveable capital [18]. Beijing has put extra efforts in creating a well-rounded prosperous society, tackling issues such as urban challenges and "big city disease" head-on [45]. The so-far progress is marked by the alleviation of non-capital functions, bolstering the capital's capabilities, and fostering a new era of coordinated development and mutual benefit among Beijing, Tianjin, and Hebei [15]. Expectations are high for Beijing to emerge as a global exemplar of urban liveability and harmony, setting precedents for other cities nationwide and worldwide.

Moreover, Beijing's leadership role extends beyond its administrative boundaries, as it takes the initiative to mitigate the strain of excessive centralisation by dispersing non-essential functions to surrounding regions. This strategic decentralisation aims to alleviate pressure on Beijing's infrastructure and resources while fostering balanced development across the broader Beijing-Tianjin-Hebei region.

3.2 Beijing and Its Ongoing 2035 Development Goals

The earlier plans indicate that the city has to build an international first-class harmonious and livable city in order to achieve significant results in the management of "big city disease", optimise the functions of the capital city, enter the forefront of the world [32] in terms of the comprehensive competitiveness of the city, and basically form the framework of the Beijing-Tianjin-Hebei world-class city cluster [15]. The envisioned progress encompasses not only physical infrastructure enhancements but also socio-economic development, environmental sustainability, and cultural enrichment. By fostering synergy and cooperation with regional cities and hubs, a new paradigm of regional development is poised to emerge, characterised by mutual benefits and shared prosperity for all stakeholders involved [14, 2, 29]. Additionally, Beijing aims to strengthen its role as a hub for international cooperation and communication while promoting balanced regional development within the broader Beijing-Tianjin-Hebei region.

3.3 Beijing and Its Long-Term 2050 Development Goals

According to Beijing Municipal Government Website [4], it is reported that Beijing aims to become the capital of a rich, strong, democratic, civilized, harmonious and beautiful modernised socialist power [42]. In another words, it is moving towards becoming a city of global influence, a model of sustainable development for mega-cities, and to build a world-class urban agglomeration with the capital city as its core. While it addresses new ideas for a good ecological environment, it also focuses on developing two aspects of economy and culture to ensure achieving a harmonious and stable society. These ambitious goals reflect Beijing's commitment to sustainable, inclusive, and harmonious development but also explore what the Chinese Dream [17] entails from the capital's perspective.

Beijing's long-term 2050 development goals outline a visionary blueprint for the city's evolution, capturing ambitious aspirations across various dimensions. Foremost among these objectives is the establishment of a sustainable, resilient, and liveable metropolis that serves as a model for urban development worldwide. Key priorities include further reducing environmental impact, enhancing green spaces, and transitioning to renewable energy sources to mitigate climate change effects.

Economically, Beijing seeks to maintain its position as a global economic powerhouse by fostering innovation, supporting entrepreneurship, and diversifying industries. Strategic investments in infrastructure, technology, and human capital are envisioned to drive sustainable economic growth and prosperity [15]. In terms of governance, Beijing aims to enhance transparency, accountability, and citizen participation, ensuring responsive and efficient administration that meets the evolving needs of its residents [21]. Furthermore, Beijing is committed to strengthening its role as a global leader in diplomacy, fostering international cooperation, and contributing to global peace and development. This role has become more visible since the start of recent conflicts or wars in various regions, and particularly between Ukraine and Russia, Saudi Arabia and Iran, and Palestine and Israel.

4 Recent Urban Development Case of Beijing Central Business District (CBD)

Beijing Central Business District (CBD) is one of the six high-end industrial functional zones in the capital, located in the central Chaoyang District of Beijing, with an area of nearly 7 km^2 [22]. It is a planned gross floor area of nearly 20 million square meters, from Chaoyang North Road in the north to the Tonghui River in the south, from Dongdaqiao Road in the west, and from the East Fourth Ring Road in the east [6]. In 1993, the concept of building a business center district was first proposed and documented in "Beijing Urban Master Plan (1991–2010)", and since then, it has entered a period of rapid development. Entering the new century, the Central Business District (CBD) has entered a phase of swift expansion, witnessing the consolidation of its industrial and spatial structures. With modern business functions progressively maturing, its international sway escalates annually, cementing its status as the capital's showcase for internationalization. Serving as a pivotal hub for high-end industries, the CBD (Fig. 6) embodies notions of efficiency and sophistication, serving as a prominent symbol of the city's identity and appeal.

The core area of CBD is in the northeast corner of Guomao Bridge, with a total land area of about 30 hectares [39]. In 2010, Beijing CBD took the development and construction of the core area as an opportunity to explore the new mechanism of docking between the planning and design, land market and industrial development. Firstly, it was meant to realise the "three bidding", i.e., integrating three bidding for the right to use the state-owned land, planning and design, and industrial projects. The second was to put forward the "three principles", that were regarded as "competing programs, competing strength, and price limit", emphasising the strongest comprehensive strength wins [3]. And this was based on a clear industrial orientation, designating the types of industries encouraged by the various plots of land, and selecting the best in line with the industrial positioning of the comprehensive assessment of the bidding enterprises. Up to now, 14 land parcels in the core area have been successfully listed and traded, attracting 41 headquarters-type enterprises and international financial institutions such as CITIC Group, etc. Also, public supporting facilities are progressing according to the overall plan, and project construction is in full swing, which was expected to bring in more than 50 billion yuan in tax revenues annually [3]. The project has and will continue to create favorable conditions for the construction of the capital city as the capital of high-end headquarters gathering.

5 The Local Government Projection for the Future Development of the City

At the city scale, Beijing is focusing on strictly controlling population size and optimising population distribution. In accordance with the requirement of determining the population by water, based on the amount of available water resources and per capita water resources, it was determined that the size of Beijing's resident population should have been controlled at less than 23 million people by 2020, and stabilised at this level in the long term after 2020.

The city also realises a reduction in the scale of urban and rural construction land. In recent years, the city has adhered to the bottom line of the scale of construction land and has strictly implemented the land-use control system. The total scale of the

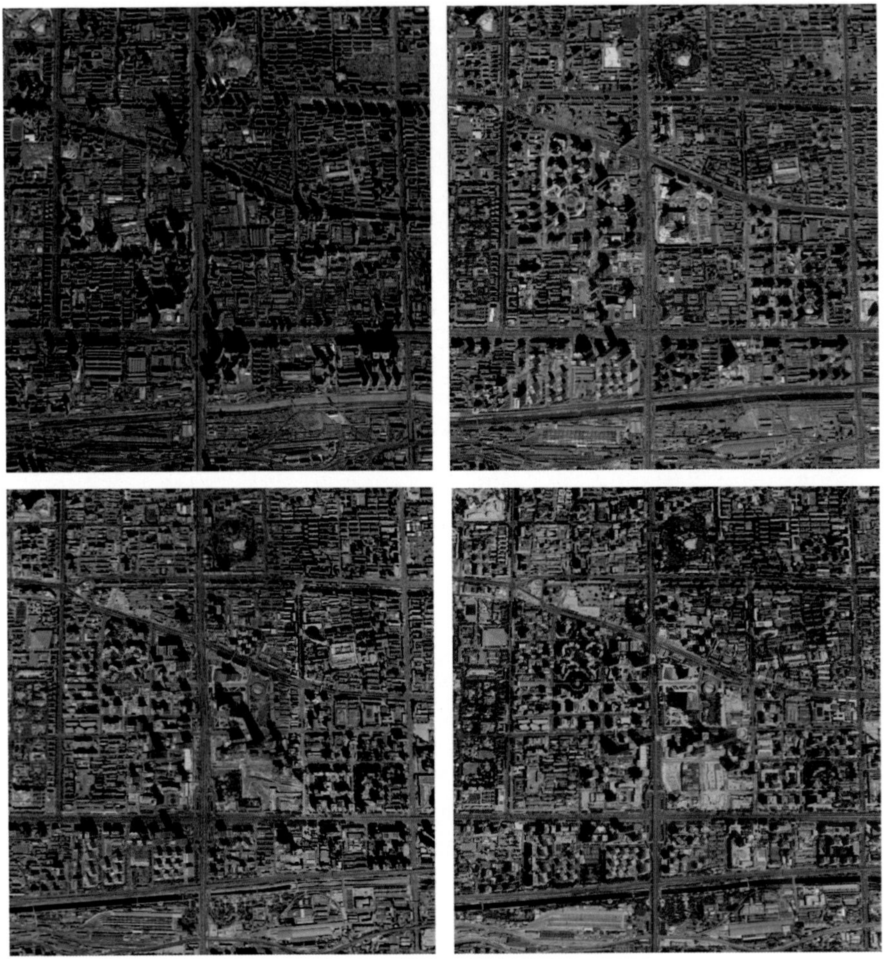

Fig. 6 Beijing's CBD Development and urban changes in the area from 2001 to 2024, with ongoing construction in south-eastern sites (maps are extracted from open and accessible maps on Google Earth, generated in April 2024)

city's construction land (including urban and rural construction land, special land, land for external transportation, and part of the land for water conservancy facilities) were controlled at less than 3720 km^2 by 2020, and will be controlled at around 3670 km^2 by 2035. By promoting the reduction of urban and rural construction land to improve quality and intensive and efficient utilisation, the scale of urban and rural construction land were reduced from the earlier status quo of 2921 km^2 to about 2860 km^2 in 2020. This continues to develop to about 2760 km^2 by 2035.

Another direction for further consideration is to reduce the intensity of development in the plains. This plan highlights three main areas of: (1) reducing the scale of urban and rural construction land in the plain area, (2) adjusting the land use structure, reasonably safeguarding land for regional transportation and municipal infrastructure and public service facilities, and (3) expanding ecological space, so that the development intensity of the plain area will be gradually reduced from the 2020 status quo of 45% to less than 44% by 2035.

The plain area of Beijing covers an area of about 6338 km^2. The key to reducing the development intensity of the plain area is to reduce the total amount and optimise its structure. Efforts are already made to compress the scale of general urban and rural construction land, appropriately increase the scale of construction land for key services and guarantees and reduce the proportion of construction land area to the area of the plains. In the short-to-midterm plan, it is anticipated that Beijing strengthens its classification guidance so that development intensity will be determined according to different regions.

5.1 Beijing's Recent and Ongoing Challenges in Sustainable Development

The development of the capital city faces many difficulties and challenges, and there are still many problems and deficiencies in the work of the Government. Mainly, it is argued that there is still a gap in the "four centers" function construction and the "four services" level compared with the party and the people's requirements. It is also evident that economic recovery is not yet a solid foundation. In this regard, enterprises, especially small and medium-sized micro-enterprises (or SMEs) have more difficulties in production and operation. There are also challenges related to high level of scientific and technological self-reliance and high precision. Hence, the combination of high-level scientific and technological self-reliance and high-precision and sharp industrial development needs to be further strengthened.

In addition, the key areas of the "bottleneck" problem still need to be broken through, such as, public health, medical care, the environment, transportation, education, pensions and other areas have a lot of shortcomings. In most cases, the problem of unbalanced and insufficient development of urban and rural areas is still prominent. For instance, the quality of the units in the government system and their style of work cannot yet be fully adapted to the needs of the new situation and new tasks [36]. The effectiveness of government services, the level of grass-roots governance and the ability to safeguard the city's emergency response in extreme situations still need to be further improved.

Other reports show that contradictions are still prominent between population, resources, environmental issues. In a way, the optimisation of the capital's functions still show signs of gradual improvement that need further institutional support. It is witnessed that sustained economic recovery is not a solid foundation and the market is expected to be weak, which has impacts on population distribution measures and strategies at the regional level. As it stands, the production and operation of enterprises is more difficult, and as a result, the residents of the consumer and private investment in the confidence is insufficient. Despite efforts to emphasis on financial and technological enterprises and services, key core technology breakthroughs in the window period has become shorter. The impact has shown that the resilience of the industrial chain supply chain and competitiveness needs to be strengthened. Also, there is still a big gap between the level of fine governance and the expectations of the masses.

One of the key areas of concern is that the management of "big city disease" requires long-term efforts. At the regional development level, the development gap between urban and rural areas is still large, and there are a lot of shortcomings and weaknesses in the field of education, medical care, pension, and other areas of people's livelihoods. The ability to prevent, mitigate and respond to disasters under extreme circumstances needs to be strengthened, and the level of production safety, fire prevention and control and emergency management needs to be improved. In all cases, we note that the ability of units and state-owned enterprises or organisations in the government system should highlight a better quality and style of work. As pe the governmental reports [37], there are still signs of deficiencies in the competence and work style of the cadres in the government system.

6 Key Lessons Learnt from Beijing's Recent and Ongoing Development Plans

Beijing's urban development strategy adeptly balances tradition and modernisation, seamlessly merging historical heritage into the fabric of a modern metropolis [34]. Through strategic urban planning, the city prioritises sustainable growth, integrating green spaces and efficient infrastructure to create a harmonious and livable environment. This approach not only preserves Beijing's unique cultural identity but also ensures long-term environmental sustainability. Here, we summarise four overarching principles or lessons learnt for.

1. **Balancing Tradition and Modernisation**: Successfully integrates historical heritage with modern urban development, creating a unique city identity;
2. **Economic Diversification**: Diversifies industries, particularly in technology and finance, promoting resilience and economic stability;
3. **Investment in Innovation**: Zhongguancun Science Park exemplifies the importance of investing in innovation hubs for technological advancement;
4. **Legacy of Event Hosting**: Demonstrates the impact of hosting major events in catalysing urban development and leaving lasting infrastructure;

To achieve these four factors should be highlighted well. First, strategic urban planning that should ideally emphasise planned growth, incorporating green spaces and efficient infrastructure for sustainable development. Second, planning paradigms should incorporate environmental sustainability measures in order to prioritise green building standards and eco-friendly initiatives, providing a model for global sustainability efforts. Third, it is important to highlight collaboration with the private sector through successful urban development strategies or mechanisms that involve collaboration between the government and the private sector. And fourth, there is a need for a more comprehensive community engagement, such as those that could promote inclusive planning and community engagement to foster a sense of ownership and social cohesion in urban development.

References

1. Acharya SK (2005) Urban development in post-reform China: Insights from Beijing. Norsk Geografisk Tidsskrift-Norw J Geogr 59(3):228–236
2. Bai H, Li N (2023) Exploring green economic efficiency trends in dominant Chinese urban agglomerations: a super-efficient SBM model approach. J Know Econ 15(3):10325–10348
3. Beijing Municipal Bureau of Statistics (2012) Introduction to the Beijing Central Business District (CBD). 北京商务中心区(CBD)简介. 北京市统计局. Available from: https://nj.tjj.beijing.gov.cn/nj/qxnj/2012/Data/cbd.htm
4. Beijing Municipal Government Website (2024) Beijing Urban Master Plan (2016–2035) 北京城市总体规划(2016年-2035年), extracted partially from https://test-axiscloud.bjaxiscloud.com.cn/weben/index.html
5. Chaolin G, Cook IG (2012) Beijing: Socialist Chinese capital and new world city. Planning Asian cities. Routledge, New York and London, pp 90–130
6. Chaoyang District People's Government of Beijing Municipality (n.d.) Institutional functions. Available from: http://www.bjchy.gov.cn/affair/govintro/depIndex_hbdw.html?depId=402883c2653629170165f5dbb8d600f7&sgs=/affair/govintro/402883c2653629170165f5dbb8d600f7
7. Chen W, Zhang S, Kong D, Zou T, Zhang Y, Cheshmehzangi A (2023) Spatio-temporal evolution and influencing factors of China's ICT service industry. Sci Rep 13(1):9703
8. Cheshmehzangi A (2016) China's New-Type Urbanisation Plan (NUP) and the foreseeing challenges for decarbonization of cities: a review. Energy Procedia 104:146–152
9. Cheshmehzangi A (ed) (2022) Green infrastructure in Chinese cities. Springer
10. Cheshmehzangi, A., & Chen, H. (2021). *China's sustainability transitions: low carbon and climate-resilient plan for carbon neutral 2060*. Singapore: Springer Nature.
11. Cheshmehzangi A, Chen H, Cheshmehzangi A, Chen H (2021a) Key suggestions and steps ahead for China's carbon neutrality plan. In: China's sustainability transitions: low carbon and climate-resilient plan for carbon neutral 2060. Springer, Singapore, pp 189–206
12. Cheshmehzangi A, Dawodu A, Sharifi A (2021b) Sustainable urbanism in China. Routledge, New York
13. Cheshmehzangi A, Xie L, Tan-Mullins M (2021c) Pioneering a Green Belt and Road Initiative (BRI) alignment between China and other members: mapping BRI's sustainability plan. Blue-Green Systems 3(1):49–61
14. Cheshmehzangi A, Tang T (2022a) China's city cluster development in the race to carbon neutrality. Springer Nature Singapore, Singapore
15. Cheshmehzangi A, Tang T (2022b) Beijing-Tianjin-Hebei city cluster: from a political center to a major industrial and innovation chain. In: China's city cluster development in the race to carbon neutrality. Springer Nature Singapore, Singapore, pp 35–46
16. da Silva MA (2014) Endless city, endless congestion: interactions between urban form and mobility in Beijing. Doctoral dissertation
17. de Sousa CR (2021) The Chinese dream: challenges and impacts for China's foreign policy. Master's thesis, ISCTE-Instituto Universitario de Lisboa (Portugal).
18. de Vries WT (2021) Urban greening for new capital cities. A meta review. Front Sustain Cities 3:670807

19. Fan J, Chen D (2013) Technological advantage of Beijing and its effect on urban development. Innovation and Regional Development in China. Routledge, New York, pp 262–292
20. Gálvez L, Cheshmehzangi A (2015) China's urban housing: the review of three housing typologies and patterns. In: Proceedings of the 6th conference on urban space and social life: Theory and practice, Macau, China, pp 6–11, June
21. Gebhardt H (ed) (2018) Urban Governance, Spatial Planning and Economic Development in the 21th Century China, vol 6. LIT Verlag Münster, Berlin
22. Gu H, Chen Y (2017) The spatial distribution characteristics of enterprises in Beijing Chaoyang CBD. In: International Conference on Transformations and Innovations in Management (ICTIM 2017). Atlantis Press, pp 762–774
23. Huang D, Liu Z, Zhao X, Zhao P (2017) Emerging polycentric megacity in China: an examination of employment subcenters and their influence on population distribution in Beijing. Cities 69:36–45
24. Jaivin L (2014) Beijing. Reaktion Books, London
25. Khan MK, Sandano IA, Pratt CB, Farid T (2018) China's Belt and Road Initiative: a global model for an evolving approach to sustainable regional development. Sustainability 10(11):4234
26. Li B (2003) An analysis of the development of Beijing CBD on the basis of CBD-system theory. J Beijing Inst. Technol (Soc Sci Ed) 6:46–48
27. Lin D, Allan A, Cui J (2015) The impact of polycentric urban development on commuting behaviour in urban China: evidence from four sub-centres of Beijing. Habitat Int 50:195–205
28. Liu X, Gao T, Wang X, Liu X, Gao T, Wang X (2018) Beijing: from high-tech to business model innovation. In: Regional innovation index of China: 2017: how frontier regions innovate, pp 111–130
29. Mangi E, Chen W, Heath T, Cheshmehzangi A (eds) (2024) Innovative public participation practices for sustainable urban regeneration. Springer, Singapore
30. Peng FL, Qiao YK, Zhao JW, Liu K, Li JC (2020) Planning and implementation of underground space in Chinese central business district (CBD): a case of Shanghai Hongqiao CBD. Tunn Undergr Space Technol 95:103176
31. People.cn (2024) Examining the Influence of Chinese automotive brands through the history of the "C-Spot" at auto shows. "从车展"C位"变迁史看中国汽车品牌影响力". Available online at: http://finance.people.com.cn/n1/2024/0504/c1004-40228717.html
32. Shao H, Lv S, Cao C (2023) The fairness and influencing factors of the spatial distribution of public services in Beijing–Tianjin–Hebei. Sustainability 15(12):9217
33. Shaffer G, Gao H (2020) A new Chinese economic order? J Int Econ Law 23(3):607–635
34. Sheth JN, Sinha M, Shah R (2016) Breakout strategies for emerging markets: Business and marketing tactics for achieving growth. FT Press
35. Teo HC, Lechner AM, Walton GW, Chan FKS, Cheshmehzangi A, Tan-Mullins M, Chan HK, Sternberg T, Campos-Arceiz A (2019) Environmental impacts of infrastructure development under the belt and road initiative. Environments 6(6):72
36. The People's Government of Beijing Municipality (2023). 2023 Annual Government Work Report. 2023年政府工作报告. 北京市人民政府网站. Available from: https://www.beijing.gov.cn/gongkai/jihua/zfgzbg/202301/t20230128_2907344.html
37. The People's Government of Beijing Municipality (2024). 2024 Annual Government Work Report. 2024 年政府工作报告. 北京市人民政府网站. Available from: https://www.beijing.gov.cn/gongkai/jihua/zfgzbg/202401/t20240129_3547363.html
38. Wang H (2010) Hi-tech industrial cluster in China-based on the case of Zhongguancun Science Park. Available online from: https://www.diva-portal.org/smash/record.jsf?pid=diva2%3A829638&dswid=-3887
39. Wang H, Gong Z, Chen Y, Liu S (2021) Research on logistics distribution management mode of Beijing CBD core area. In: The 2nd International Conference on Economics, Education and Social Research (ICEESR 2021). Available online from: https://webofproceedings.org/proceedings_series/ESSP/ICEESR%202021/ZJ_DAM20100.pdf

40. Wang J (2023) The development dilemma of the 798 art zone, Beijing, China. Drexel University, Philadelphia
41. Wonglimpiyarat J (2016) Challenges for China's banks: investment policies to support technology-based start-ups. Technol Soc 46:49–57
42. Xi J (2017). Secure a decisive victory in building a moderately prosperous society in all respects and strive for the great success of socialism with Chinese characteristics for a new era. In: Delivered at the 19th National Congress of the Communist Party of China October, vol 18, no 2017, pp 2017–2011, October
43. Xiao L, Wang C, Xu Z (2021) The role of technology business incubation in supporting innovative start-ups in China. In: Handbook of research on business and technology incubation and acceleration. Edward Elgar Publishing, pp 246–259
44. Yan X, Liu LY, Zhou S, Zhang Y (2013) The pluralization and government-planned mechanisms of the central business district in Shenzhen, China. Producer Services in China. Routledge, New York, pp 290–313
45. Yan Z, Jiang L, Huang X, Zhang L, Zhou X (2023) Intelligent urbanism with artificial intelligence in shaping tomorrow's smart cities: current developments, trends, and future directions. J Cloud Comput 12(1):179
46. Yuan J (2018) Beijing's institutional-balancing strategies: rationales, implementation and efficacy. Aust J Int Aff 72(2):110–128
47. Yusuf S, Nabeshima K (2006) China's development priorities. World Bank Publications
48. Zhang L, Yuan EJ (2023) Entrepreneurs in China's 'Silicon Valley': State-led financialization and mass entrepreneurship/innovation. Inf Commun Soc 26(2):286–303
49. Zhang P, Li J, Zhang W (2022) Characteristics of high-technology industry migration within metropolitan areas—a case study of Beijing metropolitan area. Sustainability 14(19):12750
50. Zhou Y (1998) Beijing and the development of dual central business districts. Geogr Rev 88(3):429–436

Chapter 3
Shanghai: A Vibrant Metropolis Where Modernity Meets Traditions Through Cultural Fusion

1 Introduction to Shanghai and Its Development

Shanghai is a traditional large and developed city, with the total area of 6340.5 km^2 and the built-up area of 1237.85 km^2 [19]. By 2019, Shanghai had 16 districts, including 107 subdistricts, 106 towns, and 2 townships [16]. According to the national population census in 2020, Shanghai had a permanent resident population of 24.8709 million [17], while some figures suggest a population of over 28 million by now. Alongside Beijing, Chongqing, and Tianjin, Shanghai is one of the four direct-administrated municipalities in China, which is located in the east of China, on the coast near the East China Sea [58]. Its regional importance and national status indicate Shanghai's position at the national and global level [12, 13]. It is a constituent part of the Yangtze River Delta, which consists of a group of cities around the estuary of the Yangtze River.

In ancient China, the Southern Song Dynasty (1267 AD) named Shanghai as "Shanghai Town" [28]. After 25 years, the Yuan Dynasty approved the establishment of "Shanghai County", which could be considered as the birth of the City of Shanghai [43]. Due to its special location, in the summer of 1842, Shanghai was opened as a port between China and the world after the First Opium War [49]. The current Shanghai municipal was founded on 27th May 1949. After the founding of People's Republic of China, Shanghai was designated as a municipality directly under the Central Government [38], which gave Shanghai more resources and development opportunities [57]. As one of the first international cities in either China and the world [40], Shanghai kept playing and important role in the fields of international trade and cultural communication in the past 30 years (see Fig. 1).

Shanghai, being among the pioneering cities China welcomed to the global stage, has wielded significant influence for over a century, setting it apart from many other Chinese cities [30]. Simultaneously, Shanghai provided the opportunity to other countries to enter Chinese market since the nineteenth century [50]. Nowadays, numbers of people attracted by the economic prosperity, from not only the

A. Cheshmehzangi and T. Tang, *30 Years of Urban Change in China's 10 Core Cities*, Urban Sustainability, https://doi.org/10.1007/978-981-97-8846-0_3

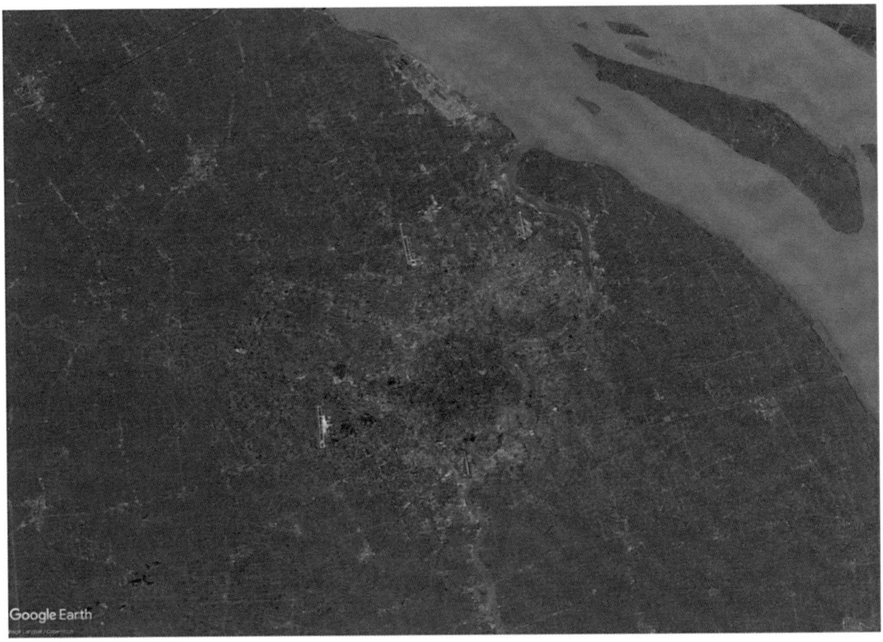

Fig. 1 Shanghai in 1985 at the start of its expansion and rapid development (map is extracted from open and accessible maps on Google Earth, generated in April 2024)

surrounding regions but also around the world (Chen, Yawei, Technische Universiteit Delft, and ProQuest, 2007). Compared with Beijing, the capital of China (see Chapter 2), Shanghai shows its higher status in the economic field. In 1992, the Chinese Communist Party (CCP) raised the goals that aimed to build Shanghai into one of the international economic, financial and trade centers. The Pudong New Area was set as the origin in the development plan, which is now one of the most developed districts in Shanghai. According to GDP of Shanghai over the years (2020), the Gross Domestic Product (GDP) of Shanghai keeps in the top 10 areas of China from 1978 to 2010. In 2017, the Chinese government [42] outlined objectives for Shanghai to achieve by 2020, including establishing the foundational structure of a globally influential science and technology innovation hub, and essentially evolving into an international hub for economics, finance, trade, shipping, and a contemporary socialist metropolis [29]. Looking ahead to 2035, Shanghai aims to transform itself into a prominent global city, renowned for its innovation, cultural richness, and ecological sustainability [36], solidifying its position as a modern international socialist metropolis with global significance [12, 48, 59, 60]. By 2050, Shanghai envisions itself as an exemplary global city, celebrated for its innovation, human-centric design, ecological consciousness, and as a modern international socialist metropolis with worldwide influence, with all development metrics meeting or surpassing international standards.

2 Shanghai's Physical Development Over the Past 30 Years

In the past 30 years, Shanghai had experienced significant changes and extensive development in almost all directions. Since the year 1964, Shanghai has kept its 10 districts and 10 counties unchanged for more than 10 years [55]. After the end of the Cultural Revolution, in 1980 and 1981, Wusong District and Minhang District were relocated [55]. In 1988, Wusong District and Baoshan County were withdrawn to Baoshan District. The pivotal year of 1990 marked a significant turning point in Shanghai's growth trajectory. It was during this time that the decision to develop Pudong was made by the CPC Central Committee and The State Council [32], ultimately transforming it into one of the most developed areas in Shanghai. In January 1993, Pudong New Area was formally established, incorporating various counties and townships into its jurisdiction. Examples include Xiepu Road Street from Yangpu District, as well as Zhoujiadu, Tangqiao, the southern wharf in the Pudong area of the south urban region, and four streets in Shanggang New Village (see Figs. 2, 3, 4, and 5).

In 1992, the government decided the abolition of Jiading County, and the establishment of Jiading District. In order to further reform and opening up, in 1993, Chuansha County, its borders and Huangpu, Yangpu, Nanshi Pudong Area were abolished. Minhang Sanlin Township merged to set up Pudong New Area. In the same year, Shanghai County and Minhang District were abolished and a new Minhang District was formed [41]. In 1995, Shanghai administrated Shanghai Pudong New Area, Xuhui, Changning, Putuo, Zhabei, Hongkou, Yangpu, Huangpu, Dinan, Shanghai

Fig. 2 Shanghai Development in 1994 (map is extracted from open and accessible maps on Google Earth, generated in April 2024)

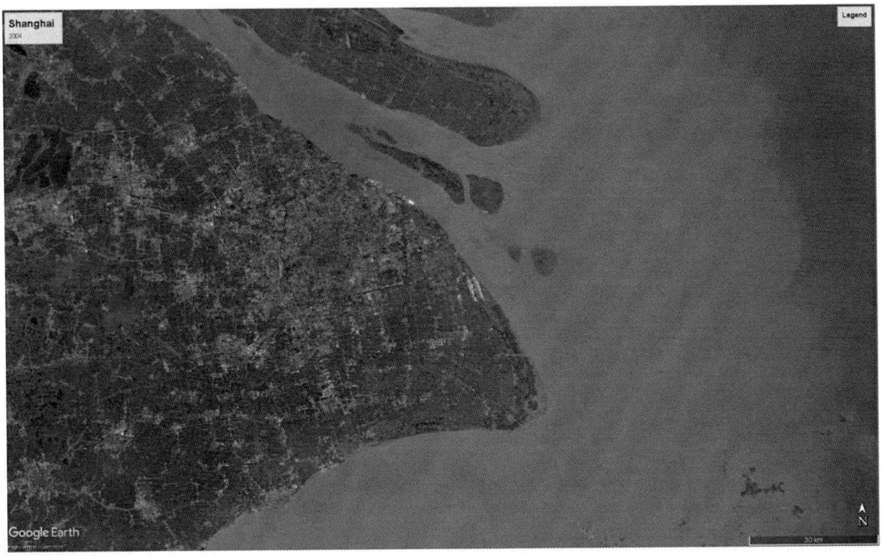

Fig. 3 Shanghai Development in 2004 (map is extracted from open and accessible maps on Google Earth, generated in April 2024)

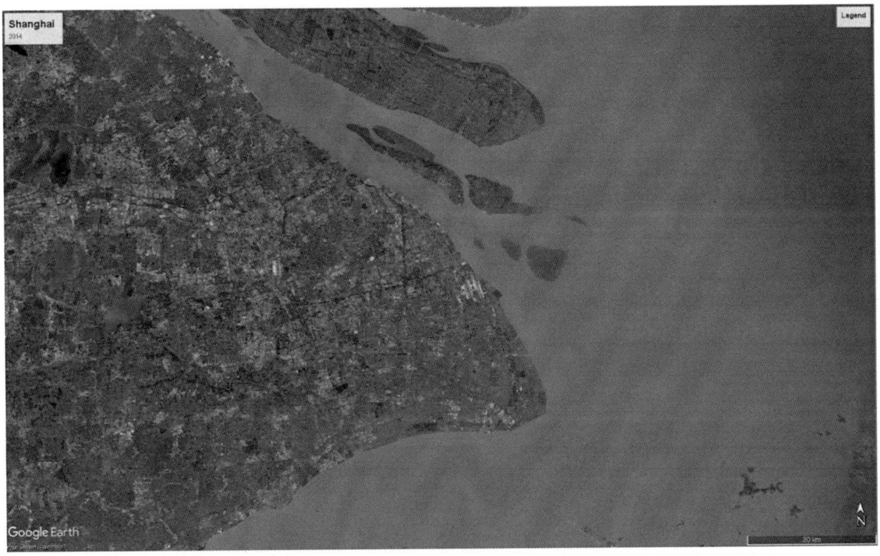

Fig. 4 Shanghai Development in 2014 (map is extracted from open and accessible maps on Google Earth, generated in April 2024)

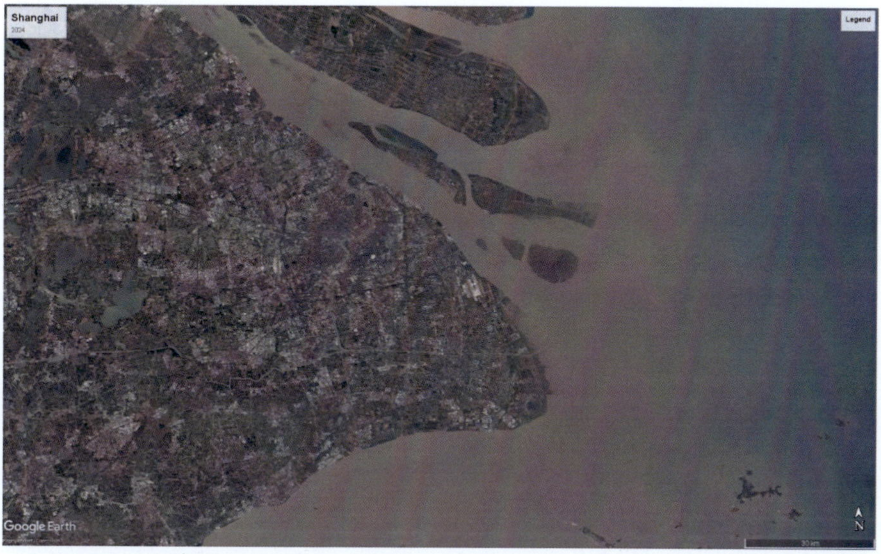

Fig. 5 Shanghai Development in 2024 (map is extracted from open and accessible maps on Google Earth, generated in April 2024)

Luwan, Jing'an, Baoshan, Minhang, Jiading and others in a total of 14 districts and Nanhui, Fengxian, Songjiang, Jinshan, Qingpu and Chongming and others in a total of 6 counties, 106 township streets, 208 towns, 10 villages, 3590 residents' committees and 2986 administrative villages. It covers an area of 6340.5 km^2, including 6218.65 km^2 of land and 697 km^2 of water [41]. At the end of the twentieth century, Songjiang County, Qingpu County and Jinshan County were also transformed into new districts of Shanghai.

In the year 2000, the south urban area was divided into Huangpu District. In January 2001, Fengxian County and Nanhui County were removed to build districts [20, 27]. In 2009, nanhui District was abolished and its administrative area was merged into Pudong New Area. In 2011, the Huangpu District and Luwan District were abolished to form a new Huangpu District [37]. In 2015, Jing'an District and Zhabei District were officially merged to form the New Jing'an District. In 2016, Chongming County under the jurisdiction of Shanghai municipality was restructured into Chongming District after being divided into districts, which marked the farewell of Shanghai municipality under the jurisdiction of counties and formed a new situation of 16 municipal districts under the jurisdiction of the whole city [1]. By June 2020, there are 107 streets, 106 towns and 2 townships in 16 districts of the Shanghai city [16].

Based on the primary economic and social metrics from the 40 years of reform and opening up [46], Shanghai's GDP surged from 31.189 billion CNY in 1980 to 3.268 trillion CNY in 2018, with the city's population rising from 16.09 million in 2000 to 24.24 million in 2018.

3 Development Strategies of Shanghai: From Opening Up to a Global Financial Center

Shanghai's development over the past three decades has been remarkable, characterised by rapid economic growth, urban transformation, and significant social progress. With the initiation of reforms and opening up in the late twentieth century, Shanghai emerged as a key driver of China's economic expansion. The decision to develop Pudong in 1990 symbolised a strategic shift towards modernisation and globalisation [35], leading to the establishment of a thriving financial, commercial, and technological hub [34]. As one of the earliest developed cites in China, the government decided to use Shanghai to realise "*the first rich to drive the second rich*". This strategy made Shanghai become an enabler of the development of other cities that along the Yangtze River [3]. Besides, in the regions that around Shanghai, the National People's Congress (NPC) of China proposed to establish regional cooperative relations which was also with the center of Shanghai.

The period spanning from 1949 to 1978 marked an era dominated by a planned economy in China. Faced with unique socio-political circumstances, the Chinese government opted for a highly centralised planned economic system modelled after the Soviet Union's approach, coupled with a strategy of closed economic development. This strategy prioritised heavy industry and emphasised self-sufficiency, with each city focusing on large-scale industrial production rather than catering to consumer demands. Consequently, cities across China underwent a transformation from consumption-oriented to production-centric entities [14, 15], resulting in a convergence of urban functions [5]. During this time, Shanghai, once a thriving international city, encountered challenges in exerting its influence on the global stage. The emphasis on heavy industry and centralised planning stifled Shanghai's traditional strengths in commerce, finance, and cultural exchange. Instead, the city became primarily geared towards industrial production, aligning with the overarching national economic agenda [24]. The highly centralised nature of the planned economy limited Shanghai's autonomy and ability to pursue its unique development path. While it remained an important economic and industrial center within China, its international influence waned as focus shifted inward towards achieving domestic production targets and economic self-sufficiency.

On the 10th of December 1982, the fifth Session of the Fifth National People's Congress of the People's Republic of China clearly proposed "*regional cooperation*" and "*preparation of the Yangtze River Delta economic zone planning with Shanghai as the center*" [54]. The meeting delineated the development strategy for Shanghai and its surrounding areas, with Shanghai as the focal point. In May 2010, it was determined that the Yangtze River Delta region comprised Jiangsu Province, Zhejiang Province, and Shanghai municipality. Building upon the framework outlined in the Plan for the Integrated Development of the Yangtze River Delta (2019), the geographic scope of the Yangtze River Delta was expanded to include Jiangsu Province, Zhejiang Province, Anhui Province, and Shanghai municipality. Earlier on, the Shanghai Municipal People's Government raised a new round of development targets from 2017 to 2020 in the master plan of Shanghai [42] with the key words of creativity, humanity and ecology. The following strategies were included in the master plan:

Strategy 1:

Shanghai's initial development strategy involves stringent population control measures. By 2020, the goal is to cap the number of permanent residents at 25 million, with continued population control as a target for 2035 [11, 33]. This is aimed at alleviating population density, employment pressure, and optimising urban spatial efficiency by dispersing overly dense populations from the city center to new towns and suburban areas. New towns are targeted to achieve population densities below 12,000 people per square kilometer, while new townships are aimed to maintain densities below 10,000 people per square kilometer.

Strategy 2:

The second strategy for Shanghai entails defining the overall quantity and composition of land designated for urban development, while promoting and directing the efficient and intensive utilisation of urban land [64]. This includes maximising the potential of existing construction sites, encouraging mixed-use development to optimise land function, and striving for a reduction in the overall area allocated for planned construction. The goal is to comprehensively enhance the efficiency of land utilisation. Additionally, there is a focus on regulating water and energy consumption to ensure it remains within sustainable limits, effectively managing resources and energy consumption [7, 8] to maintain a manageable level of environmental impact.

Strategy 3:

Shanghai's third strategic priority involves strengthening efforts in the conservation, restoration, and expansion of ecological areas. This includes intensifying collaborative measures for ecological environment management, adopting an integrated approach that encompasses both urban and rural areas [61], as well as promoting regional coordination. Additionally, there's an emphasis on enhancing biodiversity, safeguarding natural habitats, and implementing sustainable land-use practices. This strategy underscores the importance of holistic ecosystem management to ensure the long-term ecological health and resilience of the region.

Strategy 4:

Shanghai's fourth and final strategic focus entails strengthening support for critical areas such as water resources, energy, information, ecology, and national defense. This includes enhancing the resilience and security of urban infrastructure, particularly vital lifelines, to ensure their safe and reliable operation. Additionally, efforts will be made to reinforce spatial infrastructure and facilities dedicated to disaster prevention, mitigation, and relief within urban areas. Furthermore, there's a commitment to enhancing urban emergency response and recovery capabilities, emphasizing preparedness and swift action in the face of various crises or emergencies. This strategy highlights the importance of safeguarding essential services and infrastructure to sustain the city's functionality and resilience in the long term.

3.1 Regional Development Strategies for a Digital-Delta Region

The concept of the Yangtze River Delta was initially introduced on December 10, 1982 [31]. Just twelve days later, on December 22, The State Council officially released "The Notice on the Establishment of The Planning Office of Shanghai Economic Zone and Shanxi Energy Base," formally establishing the planning office for the Shanghai Economic Zone [22]. However, this office was dissolved in 1988 and replaced by the Joint Meeting of Directors of the Yangtze River Delta Cooperation Office (Committee). By 1996, the Yangtze River Delta City Economic Coordination Committee emerged as the primary coordinator for the region.

As outlined in the Regional Planning of Yangtze River Delta (2010), the geographical scope of the area initially encompassed Jiangsu Province, Zhejiang Province, and Shanghai municipality. This boundary expanded in 2019, as per the directives of the CPC Central Committee and State Council, to include Anhui Province alongside the aforementioned regions [21]. The outlined plan underscores the importance of fostering interconnected development among these areas, with a focus on five core aspects: international economy, finance, trade, shipping, and scientific and technological innovation [4]. Although there are a number of large cities in the Yangtze River Delta, the villages and counties in some places are still undeveloped. It is mentioned in the outline that government should improve the public services which could help promote the quality of rural development, for example the infrastructure connectivity between urban and rural areas [21].

The government also strategised the development of a next-generation information infrastructure, particularly focusing on enhancing the speed and capabilities of information networks through the adoption of cutting-edge 5G technologies. This initiative is expected to contribute significantly to the establishment of a Digital-Delta in the region. Moreover, ensuring a robust electrical power supply system is essential to meet the demands of such advancements. To address this need, the government has decided to accelerate the construction of several key trans-regional power projects. These include the Huainan-Nanjing-Shanghai 1000 kV UHV AC transmission channel, the Nantong-Shanghai Chongming 500 kV network project, the Shenneng Huaibei Pingshan Power Plant Phase II, and the expansion project for inter-provincial connection lines. While planning the development of the area, the government also implemented policies to help avoid the possible damage to surrounding natural environments [21], for example establishing the Xin'anjiang-Qiandao Lake ecological compensation pilot area in order to explore and establish a compensation system for wetland ecological benefits.

3.1.1 PuDong New Area: No Site Is Vacant!

In 1990, the decision to create Pudong New Area was made. Two years following this decision, the establishment of Pudong New Area received approval from the China National State Council, and subsequently, in 1993, it was officially founded. In the next two decades, a larger numbers of streets and districts were incorporated in the area [44]. As a crucial component of China's most cosmopolitan city, Pudong New Area was developed to serve as the "epicenter" within China's economic framework. Its significance was also highlighted in both the master plan of the Yangtze River Delta and The Belt and Road cooperation and initiative.

In 2019, there was a 7% increase in the GDP of Pudong New Area. Concurrently, Pudong New Area obtained municipal economic management authority within the same year [47]. On the basis of [45], the economy has achieved high-quality development on an integer platform. Gross regional product (GDP) will increase by 7%, and the total industrial output above the designated level will remain at one trillion CNY. The general public budget revenue exhibited steady growth, serving effectively as a stabilizing force, anchoring the city's economic development and acting as a key driver. Fixed asset investment, primarily fueled by industrial ventures, surged to 212.6 billion CNY. Foreign investment soared to US $8.77 billion, while total imports and exports maintained a level exceeding 2 trillion yuan, indicating sustained momentum and market vitality [10]. The financial and technology service sectors spearheaded the rapid expansion of the tertiary industry, with strategic emerging industries contributing 43.2% to the overall output value. Innovation continued to flourish, reflected in the increasing allocation of China's Belt and Road initiative spending, which reached 4.15% of GDP.

The advent of 5G, artificial intelligence, and big data technologies is fueling advancements across economic and societal domains, hastening the pace of growth in emerging industries. Concurrently, there has been a sustained enhancement in the quality of life for residents. Expenditure aimed at enhancing wellbeing rose by 9.1%, with per capita disposable income reaching 71,600 yuan. Additionally, the creation of 114,000 new jobs contributed to employment growth. The emergence of the night economy, specialised commercial complexes, and online retail platforms is reshaping urban consumption patterns. Vibrant festivals, exhibitions, and sports events add to the excitement, with total retail sales of consumer goods experiencing steady growth. Moreover, strides have been made in improving the ecological environment, with the addition of 350 hectares of green space and the planting of 13,500 mu of trees. This has resulted in a forest coverage rate of 17% and an annual air quality rate of 86.4%. Key pillars driving the development of Pudong and Shanghai include the establishment of a comprehensive financial ecosystem, enhanced trade capabilities, efficient shipping services, and robust scientific and technological innovation.

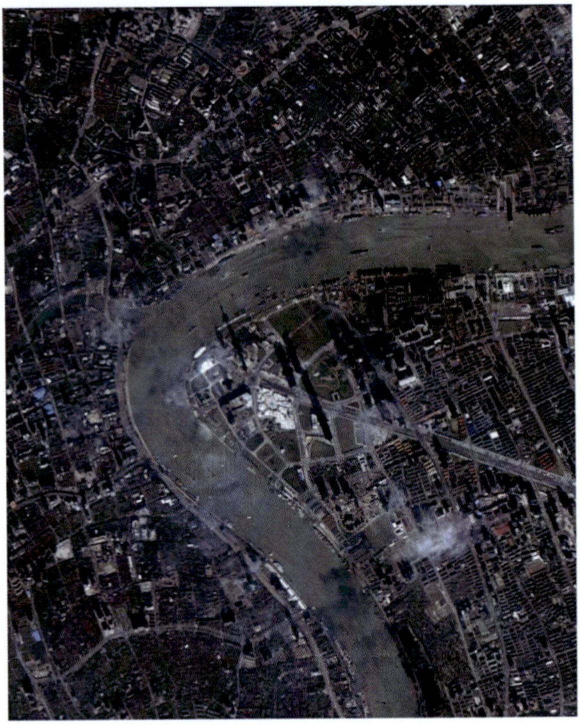

Fig. 6 PuDong's rapid development from the early days of 1990s and 2000s (in Fig. 6, map is from the year 2000) to the current highly development financial center (map is extracted from open and accessible maps on Google Earth, generated in April 2024)

The Pudong People's Government outlined its development objectives for 2020 in the overview of Pudong (2019). With 2020 marking the 30th anniversary of Pudong's development and opening up, a key focus is on establishing the fundamental framework of the core area for the science and innovation center to enhance the attractiveness, creativity, and competitiveness of cities. Efforts are also directed towards modernising urban governance to ensure management and services are more scientific, refined, and intelligent. Additionally, the government aims to implement the rural vitalisation strategy to enhance the agricultural environment, rural areas, and living standards of farmers, thereby contributing to local ecological environment construction in Pudong and Shanghai (Figs. 6 and 7).

3.2 *Shanghai's Pilot Free Trade Zone*

In August 2013, China introduced the concept of the Shanghai Pilot Free Trade Zone, which received approval from the China State Council and was officially established in September 2013 [52]. Serving as a pioneering economic experiment [39], the Shanghai Pilot Free Trade Zone designated Shanghai Waigaoqiao Free Trade Zone as its nucleus. The primary objective of the pilot free trade zone was to facilitate

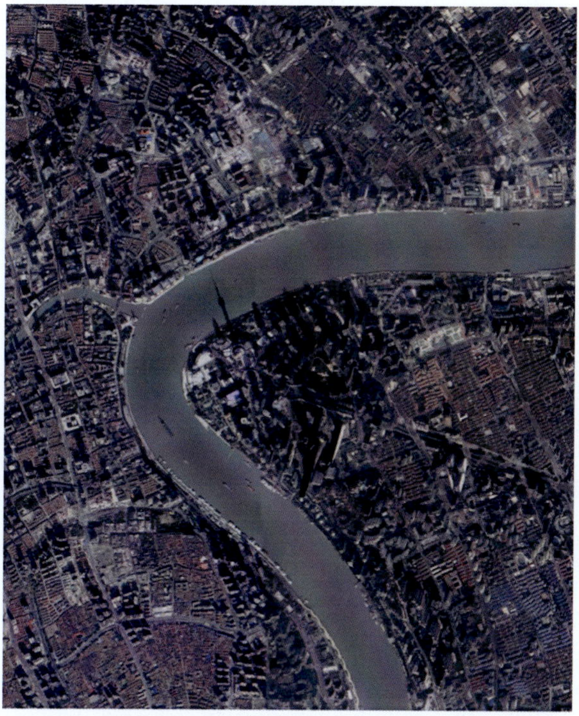

Fig. 7 PuDong's rapid development from the early days of 1990s and 2000s (in Fig. 6, map is from the year 2000) to the current highly development financial center (map is extracted from open and accessible maps on Google Earth, generated in April 2024)

the restructuring of government functions, financial systems, trade services, foreign investment, and tax policies, while fostering the growth of transit and offshore businesses in Shanghai. As outlined in the comprehensive plan for the China (Shanghai) Pilot Free Trade Zone (China National Council, 2013), the Shanghai Free Trade Zone aimed to explore the opening of the CNY under the capital account and gradually introduce financial innovations such as currency convertibility. In the forthcoming period, corporations and legal entities are expected to have the capability to conduct CNY convertibility within the Shanghai Free Trade Zone, while such privileges will not initially extend to individuals. The Shanghai Free Trade Zone is anticipated to adopt a gradual approach to yuan convertibility, potentially starting with facilitating overseas investment and financing by domestic capital. Moreover, the Shanghai Free Trade Zone is poised to play a pivotal role in the negotiations of the China-Pacific Rim Partnership (TPP) and is forecasted to serve as China's primary gateway for joining the TPP [52]. Initially, a group of 25 enterprises and 11 financial institutions have been admitted into the free trade zone. Among the financial institutions are eight Chinese banks (ICBC, Agricultural Bank of China, Bank of China, China Construction Bank, Bank of Communications, China Merchants Bank, Shanghai Pudong Development Bank, and Bank of Shanghai), two foreign banks (Citibank and DBS), and one financial leasing company (BOCOM Financial Leasing) [1].

As of November 2014, the pilot Free Trade Zone had attracted over 22,000 enterprises, inaugurated nearly 14,000 new businesses, and finalised 160 overseas investment ventures. China's outbound investment amounted to 3.8 billion USD [18]. Import clearance procedures were expedited by 41.3%, leading to a 20% rise in enterprise profitability. Furthermore, 6925 free trade accounts were established, accumulating a deposit balance of 4.89 billion CNY.

Figures 6 and 7 PuDong's rapid development from the early days of 1990s and 2000s (in Fig. 6, map is from the year 2000) to the current highly development financial center (map is extracted from open and accessible maps on Google Earth, generated in April 2024).

4 The Local Governmental Projection for Future Development Plans

According to the master plan of Shanghai 2017–2035 (2017), there are 7 main targets of Shanghai:

1. ***Improving the layout of urban space.***
   It is important to tailor actions based on the specific roles of various zones in the urban landscape and considering the available resources. Shanghai should adopt integrated planning and administration for both urban and rural regions within its city limits. Being the central city in the Yangtze River Delta, Shanghai ought to enhance collaboration and delineate responsibilities with neighbouring cities (see an example in Figs. 8, 9, and 10), laying the groundwork for establishing the Shanghai metropolitan area in the future.
2. ***Strictly controlling the size of city.***
   As a prominent global mega-city, Shanghai must commit to limiting the expansion of planned construction areas and proactively seek new approaches to its mega-city development model. By 2035, Shanghai aims to cap its permanent resident population at around 25 million and restrict the total area of construction land to no more than 3200 km^2. Additionally, emphasis will be placed on the sensible development and utilization of urban underground spaces to enhance land use efficiency.
3. ***Strengthening environmental protection.***
   Shanghai should integrate the systematic stewardship of its mountains, rivers, forests, farmlands, lakes, and grasslands, establishing a balanced urban spatial layout while fostering concurrent progress in economic, urban, rural, and environmental development. Priorities include energy conservation, emissions reduction, accelerated closure of obsolete production facilities, and rigorous regulation of overall pollutant discharge. Efforts will be intensified to comprehensively enhance the urban environment, focusing on controlling air, water, and soil pollution.

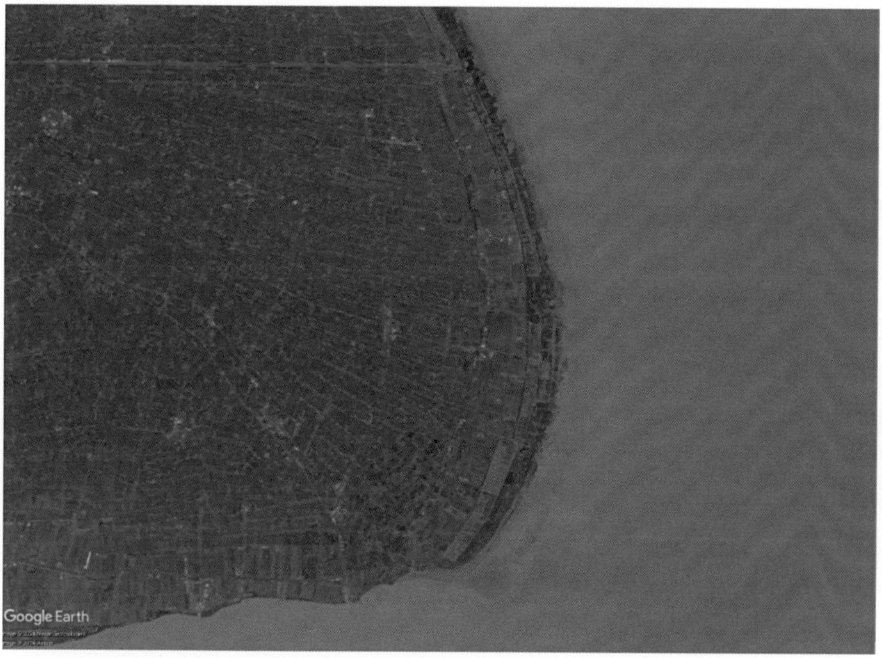

Fig. 8 Lingang area in southeast Shanghai has changed significantly from 1985 to 2004, and the map is taken from 1985 (map is extracted from open and accessible maps on Google Earth, generated in April 2024)

4. ***Creating a good living environment.***
   Striving to holistically manage crucial public service facilities such as education, culture, sports, medical care, and elderly care, recognizing their profound impact on citizens' well-being. Concurrently, enhancing the service industry's standards to elevate the quality of life, while prioritizing the creation of premium, human-centered public spaces for community enrichment and cohesion.
5. ***Shaping the characteristics of the city.***
   Remaining dedicated and adhered to upholding socialist core values, Shanghai endeavours to delve deeper into its cultural heritage, preserving historical narratives and fostering a collective memory of the city. This commitment serves to invigorate innovation and creativity within the cultural landscape, ultimately bolstering the city's soft power and allure on the global stage.
6. ***Ensuring the safe operation of the city.***
   Urban infrastructure planning and construction should align with green, circular, and low-carbon principles [14]. Enhancements to roads, railways, airports, ports, and other transportation systems are imperative, leveraging the potential of comprehensive transport hubs to enhance regional connectivity. Moreover, fortifying resilience against disasters and emergencies is crucial for effective crisis

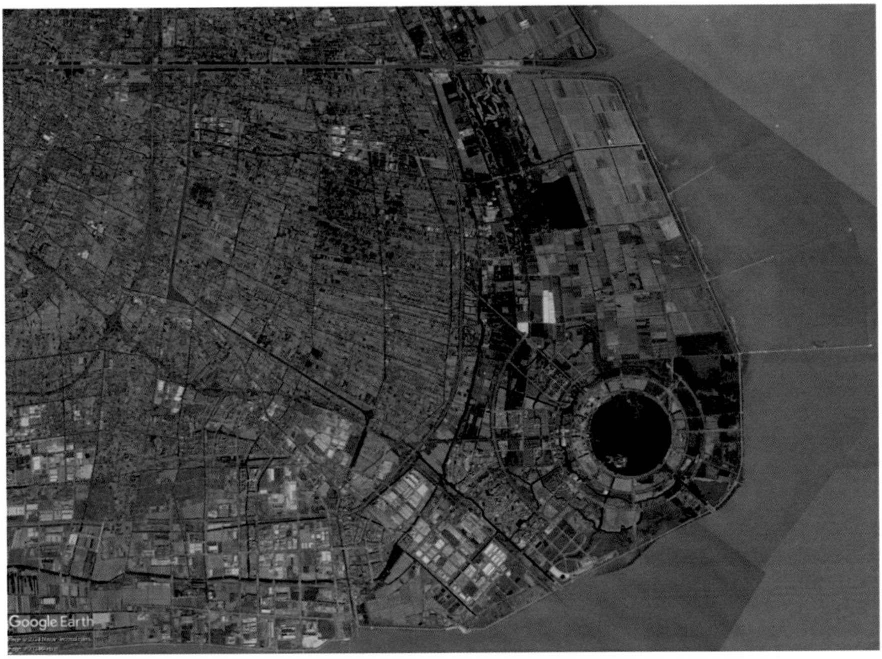

Fig. 9 Lingang area in southeast Shanghai has changed significantly from 1985 to 2004, and the map is taken from the 1990s (map is extracted from open and accessible maps on Google Earth, generated in April 2024)

management, ensuring cities are better prepared to withstand and recover from adverse events.

7. ***Improving the urban management system.***
   Elevating urban management benchmarks involves leveraging internet, big data, and other information technologies to foster increasingly scientific, sophisticated, and intelligent urban governance [9]. Priority is placed on addressing pressing issues impacting urban security, development constraints, and public concerns, thereby reinforcing comprehensive enhancements and establishing sustainable, routine management protocols.

4.1 Challenges of the Past and the Present

As a mega city, Shanghai had 2.49 million permanent resident population by 2020 (China National Bureau Statics, 2021). To transform Shanghai into a more livable metropolis, the government is committed to managing the population size within mega-cities, ensuring sustainable growth by continuously refining population demographics and spatial distribution. This entails effectively regulating population density and optimizing per capita construction land allocation to strike a

Fig. 10 Lingang area in southeast Shanghai has changed significantly from 1985 to 2004, and the map is taken from 2004 (map is extracted from open and accessible maps on Google Earth, generated in April 2024)

balance between urban development and environmental preservation. By implementing strategic measures to control population growth, such as promoting balanced migration and enhancing urban planning, Shanghai aims to foster a harmonious urban ecosystem where residents can enjoy improved quality of life, enhanced access to amenities, and a healthier living environment. Moreover, initiatives to optimize population distribution seek to alleviate congestion, enhance resource efficiency, and promote equitable access to urban infrastructure and services, contributing to the city's long-term sustainability and resilience.

In the progress of recent three to four decades, Shanghai has consumed a large number of resources, including the land resource, water resource, different kinds of energy and labor force. In the future development, the government should clarify the total amount and structure of land for urban construction, encourage and guide the economical and intensive use of land for urban construction, increase the potential of existing construction land, promote appropriate mixed use of land use functions, achieve negative growth in the total scale of planned construction land, and comprehensively improve the efficiency of land use. Effectively regulate the consumption of water resources and energy to ensure that the total consumption of resources and energy is within the bearing range. While there should be new policies that support the increase of ecological space.

Among the paramount principles lies the persistent commitment to sustain the safety standards of urban production and operations. Being a coastal mega-city, Shanghai encounters various natural hazards, including frequent occurrences of typhoons throughout the year. In response, the government is strategizing to bolster city infrastructure and resources dedicated to disaster prevention, mitigation, and relief efforts [25]. This involves enhancing spatial provisions and facilities geared towards fortifying urban resilience, as well as refining emergency response protocols and recovery mechanisms to effectively navigate through challenging situations.

As a pivotal hub in southeastern China, Shanghai is tasked with advancing its leadership role across various sectors including finance, trade, shipping, culture, and technological innovation. This entails not only strengthening Shanghai's influence but also extending its reach to bolster the development of the surrounding Shanghai region [26]. Key objectives include advancing environmental protection, optimizing industrial layouts, fostering cultural exchange, facilitating information sharing, and fostering collaboration in overseas market expansion efforts. To achieve these goals, Shanghai must leverage its central position within the metropolitan area to radiate influence and drive development [61] [23]. This involves maximizing the impact of Shanghai's transportation network to expand the reach of its 90-min commuting radius. Additionally, concerted efforts are needed to coordinate with neighbouring areas closely linked to Shanghai in terms of industry and culture, forming an interconnected urban metropolitan framework [51]. Strategies to enhance regional functional networks, devise comprehensive infrastructure plans, facilitate joint construction projects, and implement collaborative environmental governance initiatives are essential to realize a multi-faceted model of regional cooperation and governance.

Above all, the government should improve and implement planning regulations, technical standards and norms which are suitable to the new targets and could support the city development in the future. The government will launch the formulation of the Regulations of Shanghai Municipality on Spatial Planning, and the Regulations of Shanghai Municipality on Urban and Rural Planning were revised simultaneously (China National Bureau Statics, 2021) in order to ensure that space planning and management are fully incorporated into the legal track. Timely adjust policies and regulations that do not meet planning requirements.

5 Lessons Learnt from the City for Other Contexts

Throughout its developmental journey in recent decades and particularly in the last century, Shanghai's growth has been profoundly shaped by a symbiotic relationship between its historical legacy and adaptable governmental policies [63]. Initially established as a port city, Shanghai emerged as a pivotal hub for international trade and connectivity. Its evolution witnessed pivotal moments such as China's reform and opening up in the 1980s, the establishment of the Yangtze River Delta, and the creation of the China (Shanghai) Pilot Free Trade Zone. These strategic initiatives aligned seamlessly with the city's inherent characteristics, solidifying Shanghai's status as a vital global center. Despite sharing common challenges with other megacities, Shanghai's strategic investments in transportation infrastructure, including the development of multiple airports and ports, have reinforced its prominence as an international nexus for trade, transportation, and communication, further enhancing its competitive edge on the global stage.

Over the past three decades, Shanghai has witnessed an exponential surge in population, presenting multifaceted challenges such as strain on public resources and services. In response, the Shanghai government has enacted policies aimed at managing the influx of migrants from other cities. Despite these efforts, data from the [17] indicates a consistent population increase over the last decade, albeit at a slower pace. This dilemma reflects a common dilemma faced by cities poised for significant future development, grappling with the delicate balance between growth and resource allocation.

Moreover, Shanghai has strategically pivoted towards fostering a robust high-technology industry following enhancements to basic public services and traditional manufacturing sectors. Amidst this developmental trajectory, preserving the ecological landscape has emerged as a paramount concern. The city's progress in land development and industrial expansion underscores the imperative for prioritizing ecological sustainability. The efficacy of such endeavors is contingent upon the alignment of governmental policies with ecological preservation objectives. Any deviation from this harmonious balance between urban advancement and environmental protection risks undermining Shanghai's long-term ecological integrity and overall developmental trajectory. Thus, prudent policy formulation and implementation are essential to navigating the intricate interplay between urban growth and ecological stewardship in Shanghai's evolving landscape.

References

1. Baidu Baike (2021a) Baidu Baike: Shanghai. 百度百科:上海. Administrative division. 行政区划. Available from: https://baike.baidu.com/item/%E4%B8%8A%E6%B5%B7/114606?fr=aladdin#reference-[52]-18625699-wrap
2. Baidu Baike (2021b) China (Shanghai) Pilot Free Trade Zone. 中国(上海)自由贸易试验区. Available from: https://baike.baidu.com/item/%E4%B8%AD%E5%9B%BD%EF%BC%88%E4%B8%8A%E6%B5%B7%EF%BC%89%E8%87%AA%E7%94%B1%E8%B4%B8%E6%98%93%E8%AF%95%E9%AA%8C%E5%8C%BA/7272881#reference-[11]-10983365-wrap
3. Chang WY, King G (1994) Centennial climate changes and their global associations in the Yangtze River (Chang Jiang) Delta, China and subtropical Asia. Clim Res 4:95–103
4. Chen W, Zhang S, Kong D, Zou T, Zhang Y, Cheshmehzangi A (2023) Spatio-temporal evolution and influencing factors of China's ICT service industry. Sci Rep 13(1):9703
5. Chen W, Cheshmehzangi A, Mangi E, Heath T (2022) Implementations of China's new-type urbanisation: a comparative analysis between targets and practices of key elements' policies. Sustainability 14(10):6341
6. Chen Y (2007) Shanghai Pudong Urban Development in an Era of Global-local Interaction. Netherlands: IOS, 2007. Sustainable Urban Areas; 14. Web. Technische Universiteit

Delft, and ProQuest. Available from: https://ebookcentral.proquest.com/lib/nottingham/reader.action?docID=448778&query=
7. Cheshmehzangi A (2020) Towards urban energy management: a brief review of China's recent transition in urban energy demand. Low Carbon Econ 11(01):1
8. Cheshmehzangi A (2020) Evaluating the Nexus between housing and energy sectors: the comparison of urban, Peri-urban and rural housing areas in Zhuhai, China. Energy Power Eng 12(06):314–338
9. Cheshmehzangi A (ed) (2022) Green Infrastructure in Chinese Cities. Springer, Singapore
10. Cheshmehzangi A, Chen H (2021) Key suggestions and steps ahead for China's carbon neutrality plan. In: China's Sustainability Transitions: Low Carbon and Climate-Resilient Plan for Carbon Neutral 2060. Springer, Singapore, pp 189–206
11. Cheshmehzangi A, Dawodu A, Sharifi A (2021) Sustainable Urbanism in China. Routledge, New York
12. Cheshmehzangi A, Tang T (2022) Shandong Byland city cluster: from an agricultural hub to three economic circles and industrial clusters. In: China's city cluster development in the race to carbon neutrality. Springer Nature Singapore, Singapore, pp 107–118
13. Cheshmehzangi A, Tang T (2022) China's city cluster development in the race to carbon neutrality. Springer Nature, Singapore
14. Cheshmehzangi A, Tang T (2022) China's carbon neutrality plan implementation and sustainable development process review. In: China's city cluster development in the race to carbon neutrality. Springer Nature Singapore, Singapore, pp 1–23
15. Cheshmehzangi A, Tang T (2022) Yangtze River Delta City Cluster: from integrated production and services to digital technology innovation. In: China's city cluster development in the race to carbon neutrality. Springer Nature Singapore, Singapore, pp 47–62
16. China National Bureau of Statistics (2019) Zoning code for statistical purposes. 统计用区划代码>. Available from: http://www.stats.gov.cn/tjsj/tjbz/tjyqhdmhcxhfdm/2020/31/3101.html
17. China National Bureau of Statistics (2020) The 7th National Population Census Bulletin (No.3). 第七次全国人口普查公报(第三号). Available from: http://www.stats.gov.cn/tjsj/zxfb/202105/t20210510_1817179.html
18. China (Shanghai) Pilot Free Trade Zone Administration (2014) The first group interview of the Shanghai Pilot Free Trade Zone after the expansion of the zone. 上海自贸试验区扩区后首次接受集体采访 改革经验向全市辐射. Available from: http://www.china-shftz.gov.cn/NewsDetail.aspx?NID=e08cbb89-3a7d-49bc-a645-405416483be2&CID=f672f518-99a3-4789-8964-1335104906b4&MenuType=2
19. China Statistical Information Network (2019) Statistical Communique of Shanghai Municipality on National Economic and Social Development in 2019. Available from: http://www.tjcn.org/tjgb/09sh/36201_2.html
20. Chungu G (2005) Changing spatial structure of Metropolitan Shanghai 1990–2002. In Rasier S, Volkmann K (eds) Emerging patterns of the global city region: spatial changes in Johannesburg, Mumbai/Bombay, Shanghai and São Paulo, 49. Chapter 4. Available from: https://refubium.fu-berlin.de/bitstream/fub188/18105/1/AP53.pdf#page=53
21. CPC Central Committee and State Council (2019) Outline of the Plan for regional Integrated Development of the Yangtze River Delta. 长江三角洲区域一体化发展规划纲要. Available from: http://www.gov.cn/zhengce/2019-12/01/content_5457442.htm
22. Demissie A (2018) Special economic zones: Integrating African countries in China's Belt and Road initiative. In: Rethinking the silk road: China's Belt and Road Initiative and emerging Eurasian relations, pp 69–84
23. Ding L, Yang Z, Zhang J, Chen Y, Wang X (2024) Can access to urban networks promote urban development? Evidence from the Yangtze River Delta region of China. PLoS ONE 19(3):e0300199
24. Fligstein N, Zhang J (2011) A new agenda for research on the trajectory of Chinese capitalism. Manag Organ Rev 7(1):39–62
25. Fu C, Deng T, Zhang Y (2023) Urban metabolic flow in China's megacities doubled by material stock accumulation since the 21st century. NPJ Urban Sustain 3(1):52

26. Gao C, Feng Y, Wang R, Lei Z, Chen S, Tang X, Xi M (2023) 50-year urban expansion patterns in Shanghai: Analysis using impervious surface data and simulation models. Land 12(11):2065
27. He D, Ning Y (2015) Evolution of population structure and spatial distribution in shanghai since 2000. In: Urban Development Challenges, Risks and Resilience in Asian Mega Cities, pp 227–249
28. He J, He J (2014) Characterizing Shanghai's creative industries and districts. In: Creative industry districts: an analysis of dynamics, networks and implications on creative clusters in Shanghai, pp 43–78
29. Henriot C, Lu S, Aubrun C (2018) The population of Shanghai (1865–1953): a sourcebook. In: The Population of Shanghai (1865–1953). Brill
30. Howe C (ed) (1981) Shanghai: revolution and development in an Asian metropolis. Cambridge University Press, Cambridge
31. Hu C, Xu Y, Han L, Yang L, Xu G (2016) Long-term trends in daily precipitation over the Yangtze River Delta region during 1960–2012, Eastern China. Theoret Appl Climatol 125:131–147
32. Hu R, Chen W (2019) Global Shanghai remade: the rise of Pudong New Area. Routledge, New York
33. Johnston LA (2020) China's economic demography transition strategy: a population weighted approach to the economy and policy (No. 593). GLO Discussion Paper
34. Li C (2009) Reclaiming the "Head of the Dragon": Shanghai as China's center for international finance and shipping. China Leadersh Monit 28:1–17
35. Li Y (2017) A study of the unique characteristics of the development and opening-up of the Pudong New Area of Shanghai. Studies on China's Special Economic Zones, pp 143–157
36. Liu B, Zhang Q, Liu X (2023) Research on the cultivation mechanism of young scientific and technological innovation talents in Shanghai for 2035. In: 3rd International Conference on Management Science and Software Engineering (ICMSSE 2023). Atlantis Press, pp 1001–1014, October
37. Ma QP (2019) What drives China's economy: economic, socio-political, historical and cultural factors. Routledge, New York
38. MacPherson KL (1990) Designing China's urban future: the Greater Shanghai plan, 1927–1937. Plan Perspect 5(1):39–62
39. Raven R, Sengers F, Spaeth P, Xie L, Cheshmehzangi A, De Jong M (2019) Urban experimentation and institutional arrangements. Eur Plan Stud 27(2):258–281
40. Shao Q (2013) Shanghai gone: Domicide and defiance in a Chinese megacity. Rowman & Littlefield, Lanham
41. Shanghai Local Chronicles Office (2014) The first volume: Establishing evolution. 第一卷:建置沿革. Available from: http://www.shtong.gov.cn/node2/node2247/node4562/index.html
42. Shanghai Municipal Peoples Government (2017) Master plan of Shanghai (2017–2035). 上海城市总体规划 (2017–2035). Available from: https://www.shanghai.gov.cn/newshanghai/xgkfj/2035002.pdf
43. Shanghai Municipal Peoples Government (2019) The establish of Shanghai. 上海建城. Available from: https://www.shanghai.gov.cn/nw2/nw2314/nw3766/nw3767/nw3769/u1aw1.html
44. Shanghai Pudong New Area People's Government (2017) Pudong yearbook 2017. 浦东年鉴2017. Available from: http://www.pudong.gov.cn/shpd/about/20180319/008006031003_8ba64629-dc34-40dc-a6e8-5253938ab3d6.htm
45. Shanghai Pudong New Area People's Government (2021) The overview of Pudong. 浦东概况. Available from: http://www.pudong.gov.cn/shpd/about/20200214/008001001_e32687d9-589d-4285-8411-210c7845a4f0.htm
46. Shanghai Statistics Bureau (2019) The main economic and social indicators of the 40 years of reform and opening up. 改革开放四十年经济社会主要指标. Available from: http://www.stats-sh.gov.cn/tjnj/nj18.htm?d1=2018tjnj/C0112.htm
47. Sina Finance (2019) The Pudong New Area has been given municipal economic management authority. 加大放权力度,浦东新区被赋予市级经济管理权限. Available from: http://finance.sina.com.cn/roll/2019-06-25/doc-ihytcerk9182796.shtml

48. Song E (1996) Shanghai: transformation and modernization under China's open policy. Chinese University Press, Beijing
49. South China Morning Post (2016) The Opium War (or how Hong Kong began). Available from: https://www.scmp.com/article/974360/opium-war-or-how-hong-kong-began
50. Tan-Mullins M, Cheshmehzangi A, Chien S, Xie L (2017) Smart-eco cities in China: trends and city profiles 2016. Exeter: University of Exeter (SMART-ECO Project)
51. Tao D, Zhou W (2022) An evaluation and optimization of green development strategy for the Nanjing-Hangzhou Eco-Economic Zone in China. Sustainability 14(24):17002
52. Tencent Finance (2013) Shanghai free Trade Zone will open, which will benefit ports, airports, warehousing, real estate and finance. 上海自贸区将开放 港口机场仓储地产及金融受益. Available from: https://finance.qq.com/a/20130704/001531.htm
53. The Central People's Government of the People's Republic of China (2007) The New mission of the Yangtze River Delta: The Historical orientation of China's regional economy from the Yangtze River Delta. 长三角新使命-从长三角看中国区域经济历史定位. Available from: http://www.gov.cn/jrzg/2007-10/31/content_791609.htm
54. The Paper (2020) As a world-class city cluster, the Yangtze River Delta will contribute new culture and civilization to the world. 长三角成为世界级城市群,要为全球贡献新的文化与文明. Available from: https://www.thepaper.cn/newsDetail_forward_10248247
55. Wei YD, Leung CK (2005) Development zones, foreign investment, and global city formation in Shanghai. Growth Chang 36(1):16–40
56. Wenku (2020) GDP of Shanghai over the years. 上海历年GDP. Available from: https://wenku.baidu.com/view/fb199c959cc3d5bbfd0a79563c1ec5da50e2d63e.html
57. Wu W, Yusuf S (2004) Shanghai: remaking China's future global city. In: World cities beyond the west: globalization, development, and inequality, pp 27–58
58. Xiao H, Xu Z, Ren J, Zhou Y, Lin R, Bao S, Zhang L, Lu S, Lee CK, Liu J (2022) Navigating Chinese cities to achieve sustainable development goals by 2030. The Innovation 3(5):100288
59. Xie L, Cheshmehzangi A, Tan-Mullins M, Flynn A, Heath T (2020) Urban entrepreneurialism and sustainable development: A comparative analysis of Chinese eco-developments. J Urban Technol 27(1):3–26
60. Xie L, Flynn A, Tan-Mullins M, Cheshmehzangi A (2019) The making and remaking of ecological space in China: the political ecology of Chongming Eco-Island. Polit Geogr 69:89–102
61. Zhang S, Zhao J, Jiang Y, Cheshmehzangi A, Zhou W (2023a) Assessing the rural–urban transition of China during 1980–2020 from a coordination perspective. Land 12(6):1175
62. Zhang X, Ma W, Sheng S (2023b) Understanding the structure and determinants of economic linkage network: The case of three major city clusters in Yangtze River Economic belt. Front Environ Sci 10:1073395
63. Zheng E, Xue C, Chen G, Zhang Y, Zou J (2023) Unveiling urban marathon development characteristics and urban growth strategies in China: Insights from time series analysis of Baidu Search Index. PLoS ONE 18(6):e0287760
64. Zhu J (2020) Pro-growth planning regime and housing outcomes in urbanising China: A case study of Shanghai. Doctoral dissertation, UNSW Sydney

Chapter 4
Shenzhen: A Modernity Through the Promotion of Innovation, Technology, and Education

1 General information and introduction to Shenzhen

Shenzhen's remarkable transformation from a humble fishing village cluster to a global technological and economic powerhouse represents China's rapid development, particularly after the opening up era [19]. Situated in the southern part of China, Shenzhen's rise is a testament to the nation's strategic vision and commitment to progress [61]. The city's journey began modestly, with its origins rooted in a cluster of small fishing communities or villages scattered in the area (see Figs. 1, 2, 3, 4, and 5). However, fuelled by favourable government policies and strategic investments [14], Shenzhen rapidly evolved into a vibrant metropolis. Its proximity to Hong Kong, one of the world's leading financial centers, further catalysed its growth, facilitating cross-border trade and investment [18]. As the second largest city in the southern province of Guangdong, Shenzhen [9] has become a major anchor point connecting China to its non-mainland regions through vast networks of trade and business.

Primarily, Shenzhen's development has been driven by innovation and entrepreneurship [56], which echoes very well in how the city has developed in recent years (if not decades). The city possesses a flourishing ecosystem of technology companies, startups, and research institutions [21]. In recent decade or so, such factors and sectors have been a more tangible part of the city's image for development and growth [53]. From telecommunications giants to cutting-edge biotech firms, Shenzhen is now home to a diverse array of enterprises at the forefront of their respective industries. This was not the case just around 20 years ago, but it has become a tangible view of how Shenzhen is seen and experienced in the current age.

Furthermore, Shenzhen's dedication to sustainability goes beyond mere rhetoric, with concrete actions supporting its environmental objectives [55]. The city has implemented stringent regulations to reduce carbon emissions and promote energy efficiency [10]. Initiatives such as waste management programs and green

A. Cheshmehzangi and T. Tang, *30 Years of Urban Change in China's 10 Core Cities*, Urban Sustainability, https://doi.org/10.1007/978-981-97-8846-0_4

Fig. 1 Shenzhen urban areas in 1984 (map is extracted from open and accessible maps on Google Earth, generated in April 2024)

Fig. 2 Shenzhen urban areas in 1994 (map is extracted from open and accessible maps on Google Earth, generated in April 2024)

building standards reflect Shenzhen's holistic approach to environmental stewardship, ensuring a greener, more liveable city for generations to come. In addition to its economic and environmental endeavours [41], Shenzhen is also renowned for its vibrant cultural scene and architectural marvels. The city's skyline is adorned with towering skyscrapers, innovative design concepts, and iconic landmarks, reflecting

Fig. 3 Shenzhen urban areas in 2004 (map is extracted from open and accessible maps on Google Earth, generated in April 2024)

Fig. 4 Shenzhen urban areas in 2014 (map is extracted from open and accessible maps on Google Earth, generated in April 2024)

its modernity and dynamism. Some used to argue against the fact that Shenzhen is a city without identity [6]. However, we always argue that identity is relative and a temporal factor, which indicates that a city like Shenzhen has reinvented itself in a direction that can now help distinguish it from other cities.

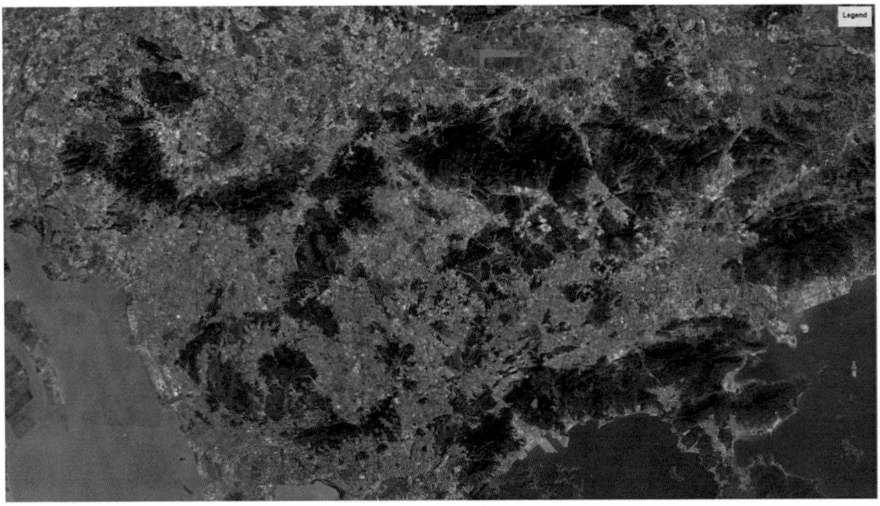

Fig. 5 Shenzhen urban areas in 2024 (map is extracted from open and accessible maps on Google Earth, generated in April 2024)

As Shenzhen continues to mark new frontiers, its journey serves as an inspiration to cities worldwide, demonstrating the transformative power of vision, innovation, and relentless determination. In less than just three to four decades of development, Shenzhen's evolution is a testament to the boundless potential of human endeavour.

Situated in the southern part of China, adjacent to Hong Kong, Shenzhen holds strategic importance as a Special Economic Zone (SEZ), playing a pivotal role in the country's economic reforms [59]. This city is recognised for its leading position in technology and commerce [60], home to numerous tech giants and startups that contribute significantly to China's global competitiveness [25]. Renowned as a beacon of technological advancement and commercial prowess [47], Shenzhen is teeming with a myriad of tech giants and burgeoning startups [29], driving China's ascension on the global stage [45]. The unique blend of its historical roots, economic significance, and technological prowess makes Shenzhen a key player in the contemporary landscape of Chinese cities [2]. This convergence also accentuates Shenzhen's indispensability in shaping the contemporary narrative of Chinese urbanisation (see Figs. 6, 7, 8, 9, 10, and 11)—particularly that is now focused on high-quality urbanisation.

Fig. 6 Part of Shenzhen's Bao'an coastal areas with the emergence of Shenzhen Bao'An International Airport, from records of 1970 to 2024 (maps are extracted from open and accessible maps on Google Earth, generated in April 2024)

2 Shenzhen's Foremost Development Over the Past 30 years

The upgraded city infrastructure in Shenzhen has significantly enhanced its functional capacity [4]; i.e., its functional quality of a modernised city has been greatly, too. The city has expedited its high-quality integrated development, eliminating all 15 "second-line passes" and effectively implementing the "enclave" development model in the Shenzhen-Shantou Special Cooperation Zone [43]. Progress in key construction projects such as the Shenzhen Bay Super Headquarters Base has accelerated, while new landmarks like the International Convention and Exhibition Center and Ping An Financial Center are emerging rapidly (see Figs. 12, 13, 14, and 15). Shenzhen has been selected as one of the initial pilot projects for the National Transportation Powerhouse initiative, with the metro network exceeding 400 km and the commencement of construction on three airport runways and the Shenzhen-Zhongshan Corridor. The expansion of the highway network is also underway. Additionally, Shenzhen leads nationally in online government service capacity and was the first globally to achieve citywide coverage of 5G independent group networks. Significant strides have been made in optimizing land use efficiency, revitalizing 93 square kilometres of underutilized land. By embracing cutting-edge technologies

Fig. 7 Part of Shenzhen's Bao'an coastal areas with the emergence of Shenzhen Bao'An International Airport, from records of 1970 to 2024 (maps are extracted from open and accessible maps on Google Earth, generated in April 2024)

such as 5G networks, Shenzhen continues to innovate and set global benchmarks [46], solidifying its reputation as a dynamic and forward-thinking city at the forefront of China's development. The city is actively promoting initiatives like the "toilet revolution" and household garbage classification [43], resulting in notable improvements in urban village aesthetics and governance standards.

Shenzhen stands out as a distinctive exemplar of urban advancement within China, chiefly due to its rapid pace of development and steep learning curves. Since the conception of Deng Xiaoping's vision for Shenzhen, the city has experienced exponential growth, focusing on pioneering mechanisms for both national and international trade and commerce. Within just under three decades, Shenzhen has escalated its developmental standards to unprecedented heights [43]. Its comprehensive economic vigor and innovation prowess have surged significantly, evidenced by an average annual GDP growth rate of 7.1% during the 13th Five-Year Plan period. Shenzhen consistently ranks among the top cities in mainland China in terms of both average and per capita GDP, while simultaneously boasting remarkably low energy and water consumption per unit of GDP compared to other large and medium-sized cities nationwide [43]. The city's industrial output value has soared to claim the top spot in China, with strategic emerging industries contributing over one trillion yuan in added value, and its high-tech sector emerging as a national flagship [20].

Fig. 8 Part of Shenzhen's Bao'an coastal areas with the emergence of Shenzhen Bao'An International Airport, from records of 1970 to 2024 (maps are extracted from open and accessible maps on Google Earth, generated in April 2024)

Shenzhen has solidified its position as the focal point of the comprehensive national science center in the Greater Bay Area, attracting prestigious national laboratories. Concurrently, major projects such as Guangming Science City, the Shenzhen-Hong Kong Science and Technology Innovation Cooperation Zone in the Loop, and Xili Lake International Science and Education City have been expedited [16]. The city's commitment to research and development is evident, with society's overall R&D investment accounting for 4.93% of regional GDP [11]. Notably, Shenzhen has pioneered the establishment of the first angel investment guide fund, amounting to a substantial 10 billion yuan, and has meticulously constructed an entire innovation ecosystem encompassing fundamental research, technology development, achievement commercialization, science and technology financing, and talent cultivation [43].

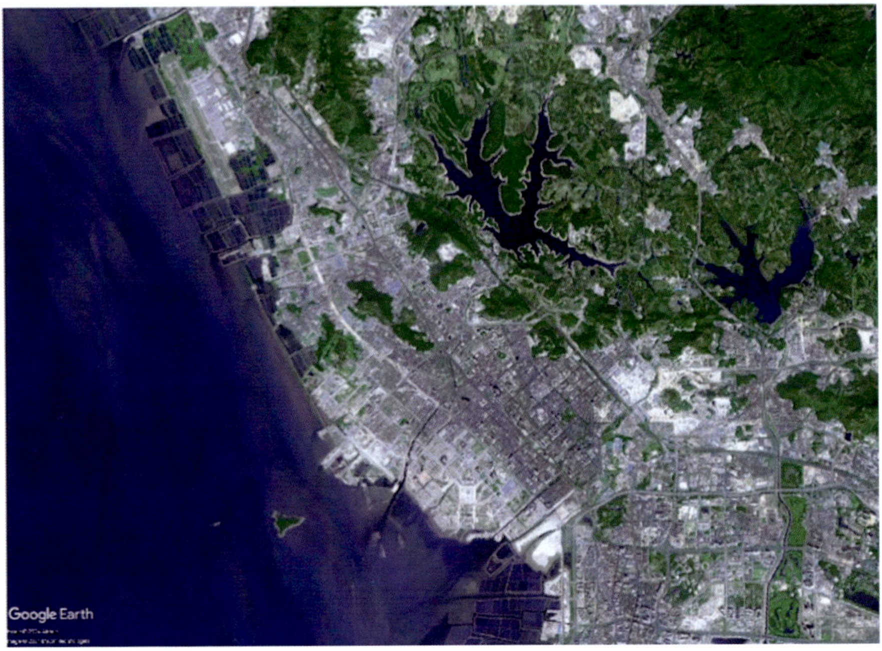

Fig. 9 Part of Shenzhen's Bao'an coastal areas with the emergence of Shenzhen Bao'An International Airport, from records of 1970 to 2024 (maps are extracted from open and accessible maps on Google Earth, generated in April 2024)

3 Shenzhen's Development Strategies and Directions: No Longer a Cluster of Fishing Villages in the South

Deng Xiaoping's vision for the Shenzhen plan epitomised his commitment to economic reform and opening up of China to the world [54]. He saw Shenzhen as a pioneering experiment, a laboratory for testing market-oriented policies and attracting foreign investment. Deng believed that by designating Shenzhen as a Special Economic Zone (SEZ), with preferential policies and flexible regulations, it could serve as a catalyst for modernisation and economic development [40]. Much of what he believed in and proposed happened over these few decades of development in Shenzhen [48]. He emphasised the importance of embracing innovation and entrepreneurship [35], envisioning Shenzhen as a beacon of progress and prosperity that would inspire similar initiatives across China, ultimately driving the nation's transformation into a global economic powerhouse.

The ongoing efforts are directed towards establishing Shenzhen as a beacon of high-quality development, exemplifying the rule of law, urban civilization, and the well-being of its residents [30]. The city aspires to serve as a model of excellence in these aspects while also pioneering sustainable development practices. These goals represent a strategic vision aimed at positioning Shenzhen as a leader in

Fig. 10 Part of Shenzhen's Bao'an coastal areas with the emergence of Shenzhen Bao'An International Airport, from records of 1970 to 2024 (maps are extracted from open and accessible maps on Google Earth, generated in April 2024)

various domains [26]. The city's persistent pursuit of progress encompasses not only economic growth but also the enhancement of legal frameworks, civic life, and overall societal welfare. Through these accomplishments, Shenzhen aims to set new standards and encourage emulation both nationally and internationally.

By 2025, Shenzhen aims to evolve into a modernized, globally-oriented, and innovative urban center, marking a significant milestone in achieving socialist modernization at its core [22]. This entails consolidating its economic prowess and advancing the quality of development to position itself among the forefront of global cities. Aligned with this vision, Shenzhen places a strong emphasis on bolstering its innovation capacity, particularly evident in the rapid advancements within the high-tech sector [62] in recent years [51]. Noteworthy achievements include the establishment of a comprehensive national science center in the Greater Bay Area [52], yielding remarkable outcomes [37]. Consequently, Shenzhen's robust investment in research and development, along with its industrial innovation capacity, ranks among the world's leading, with R&D investment accounting for approximately 5% of regional GDP. This commitment has resulted in significant enhancements in original innovation capacity and pivotal breakthroughs in key core technologies.

Fig. 11 Part of Shenzhen's Bao'an coastal areas with the emergence of Shenzhen Bao'An International Airport, from records of 1970 to 2024 (maps are extracted from open and accessible maps on Google Earth, generated in April 2024)

Moreover, Shenzhen has witnessed and continues to undergo significant enhancements in its cultural influence. It has experienced and continue to experience substantial improvement in cultural soft power [49]. The city's cultural attributes, characterised by openness, diversity, compatibility, innovation, and contemporary flair, have become increasingly prominent [7, 15]. These cultural traits have become more pronounced in recent years, with city officials advocating for a comprehensive vision that underscores the deep integration of socialist core values into the fabric of society [38]. Priority is placed on enhancing the city's level of civilization, expanding public cultural services, and fostering the development of the cultural industry. In recent times, several iconic cultural landmarks have been erected, signalling a concerted effort towards rebranding the city through its cultural assets and mechanisms with a more globally resonant impact, thus shaping its future trajectory [36]. This overarching vision is in harmony with Shenzhen's ongoing endeavours to elevate the well-being of its residents to unprecedented levels.

Finally, according to government reports, Shenzhen has achieved fundamental modernisation of its urban governance system and capacity [39]. This entails a significant enhancement in the rule of law, scientific decision-making, refinement, and intelligence in urban governance. This achievement marks a pivotal milestone in Shenzhen's evolution, underlining its commitment to progressive governance practices

Fig. 12 Rapid pace of Futian District towards becoming the main financial center and district (maps are extracted from open and accessible maps on Google Earth, generated in April 2024)

Fig. 13 Rapid pace of Futian District towards becoming the main financial center and district (maps are extracted from open and accessible maps on Google Earth, generated in April 2024)

Fig. 14 Rapid pace of Futian District towards becoming the main financial center and district (maps are extracted from open and accessible maps on Google Earth, generated in April 2024)

Fig. 15 Rapid pace of Futian District towards becoming the main financial center and district (maps are extracted from open and accessible maps on Google Earth, generated in April 2024)

Fig. 16 Rapid pace of new port region in southwestern regions (maps are extracted from open and accessible maps on Google Earth, generated in April 2024)

[42]. As a result, Shenzhen is cultivating a stable, fair, transparent, and predictable international business environment [28], positioning itself as a global leader in governance standards [43]. This transformation is gradually materialising, inducing the city towards ambitious objectives of enhanced liveability, resilience, and intelligence (see Figs. 16, 17, 18, 19, and 20). The ultimate aspiration is for Shenzhen to emerge as a paradigm for modernized governance in mega-cities [31], setting a benchmark for effective and progressive urban management worldwide.

4 Shenzhen's Central Role in the Development of the Greater Bay Area

The evolution of the Guangdong-Hong Kong-Macao Greater Bay Area entails the synchronised advancement of its 11 constituent cities and the establishment of a network of interconnected urban hubs within the Bay Area [1]. Each city within the Bay Area possesses unique functions and strategic positioning, with distinct trajectories for future development shaped by their individual strengths and weaknesses. This diversity in functions and development orientations reflects the nuanced dynamics of each city, underscored by their varying economic specialties, infrastructural capacities, and cultural identities. As a result, the Greater Bay Area emerges as

Fig. 17 Rapid pace of new port region in southwestern regions (maps are extracted from open and accessible maps on Google Earth, generated in April 2024)

Fig. 18 Rapid pace of new port region in southwestern regions (maps are extracted from open and accessible maps on Google Earth, generated in April 2024)

Fig. 19 Rapid pace of new port region in southwestern regions (maps are extracted from open and accessible maps on Google Earth, generated in April 2024)

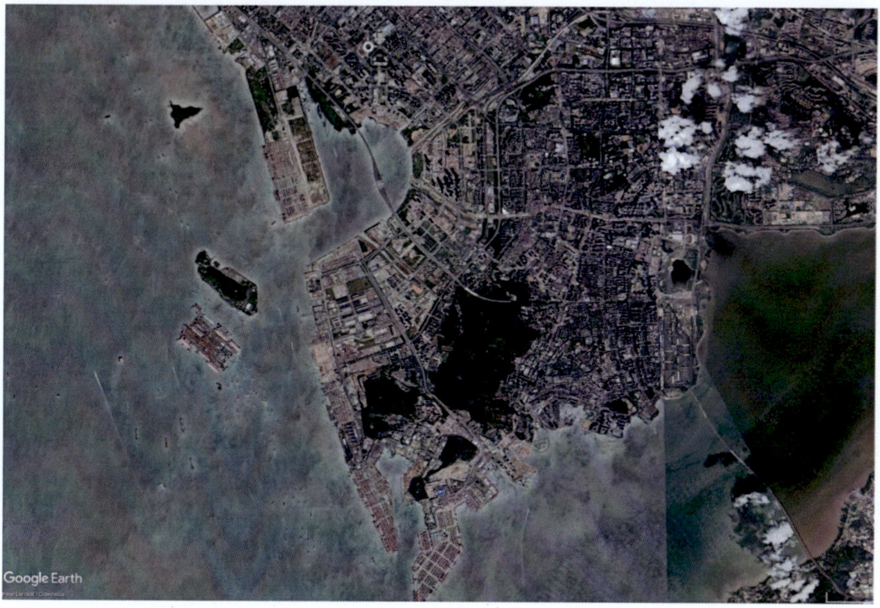

Fig. 20 Rapid pace of new port region in southwestern regions (maps are extracted from open and accessible maps on Google Earth, generated in April 2024)

a dynamic ecosystem where cities collaborate and thrive, leveraging their respective advantages to foster sustainable growth and prosperity. Shenzhen's role as a link in this strategic regional development plans means that it is the key contact point and regional node for the 11 cities in the Greater Bay Area. It is well represented as the "throat" connecting Hong Kong, Macao, and the other eight cities in the Greater Bay Area, and the hub connecting the domestic and overseas markets in this strategic region.

Shenzhen stands at the forefront of driving the development and integration of the Greater Bay Area, playing ca critical role in its economic vitality and regional synergy. As one of the key cities within this important region, Shenzhen's strategic location, robust infrastructure, and dynamic economy position it as a hub for the region's growth trajectory. At the heart of Shenzhen's influence lies its status as a global hub for innovation and technology. The city's thriving ecosystem of tech companies, startups, and research institutions fosters cutting-edge developments in fields like artificial intelligence (AI), biotechnology, and telecommunications [34]. This technological expertise not only fuels Shenzhen's own economic engine but also radiates throughout the Greater Bay Area, attracting talent and investment from across the globe. This is because the Greater Bay Area serves as a gateway for international trade and commerce, facilitated by its well-established transportation networks and modern logistics infrastructure. The decisive role that Shenzhen plays in this region is influential based on its commitment to sustainable development and urban innovation that sets a precedent for the entire Greater Bay Area [5, 6]. The city's initiatives in green technology, urban planning, and environmental conservation serve as a model for other cities in the region, driving collective efforts towards sustainable development.

Shenzhen's dominant role as a hub is primarily evident in its institutional framework, external economic interactions, market orientation, and internationalisation efforts, as well as in its public administration [34]. Hong Kong and Macao have fully embraced market economy principles, and among the cities in China and the Greater Bay Area, Shenzhen features one of the highest levels of marketisation. Shenzhen can further assimilate and adapt the efficient market economy models of Hong Kong and Macao, subsequently disseminating these practices to other cities in the Greater Bay Area. Acting as a channel for market economy systems, Shenzhen can enhance the overall marketisation of the Greater Bay Area's economy. Moreover, drawing from Hong Kong and Macao's globally acclaimed public management and service practices, Shenzhen can incorporate and transfer replicable public management mechanisms to other cities in the Greater Bay Area. Furthermore, cities in the Greater Bay Area can leverage Shenzhen's strengths to bolster their external economic engagements and enhance their urban internationalisation efforts [34], thereby leveraging their individual urban strengths for collective advancement.

4.1 Local Government Projection for the Future Development

The governmental reports provide a comprehensive roadmap for Shenzhen's future development, outlining nine key areas that will shape the city's trajectory in the coming years [23]. These areas encompass a diverse range of sectors, reflecting Shenzhen's multifaceted approach to sustainable growth and innovation [44]. Aligned with its ongoing urban construction measures and infrastructure development lay a great capacity building opportunity for innovation and technology [27]. Such mindset has developed opportunities for cultivating new talents and education development in Shenzhen. Aligned with all these efforts sit the city's overarching plan for international cooperation and exchange initiatives aim to strengthen the city's global connections and promote collaboration with international partners [50]. The city seeks to expand international trade and investment opportunities, enhance cultural exchange, and foster diplomatic relations to support its role as a global city and contribute to global development efforts. The nine key areas for future development, according to Shenzhen Municipal People's Government [44], are the following:

1. Promoting the development of the Greater Bay Area of Guangdong-Hong Kong-Macao to a new phase, comprehensively deepening reforms at various levels, and expanding high-level international openness.
2. Prioritizing the realization of new industrialization as a central objective, striving to establish a modern industrial system with heightened global competitiveness.
3. Expediting the establishment of a globally significant industrial science and technology innovation hub through intensified efforts in education, science, technology, and talent cultivation.
4. Strengthening the implementation of the strategy to boost domestic demand and actively engaging in and supporting the establishment of a unified national market.
5. Propelling the "High-Quality Development Project" for a hundred counties, thousand towns, and ten thousand villages to elevate regional development coordination to a superior level.
6. Emphasizing the development of cities as communities built by and for the people, and hastening the transition of mega-city development modes.
7. Crafting a modernized model for Shenzhen characterized by harmonious coexistence between humans and nature, underpinning high-quality development with an exceptional ecological environment.
8. Continuously enhancing the city's cultural influence and soft power, actively fostering a significant showcase of China's modern civilization.
9. Upholding the improvement of people's livelihoods as a core development principle, striving to set a standard for well-being and contentment in prosperity-building, governance, and shared development.

5 Shenzhen and Its Challenges of the Recent Past and Present

There remains a disparity between the developmental quality and efficiency of Shenzhen compared to leading international cities. Its capacity to access global resources is insufficient, and the business operational costs persistently high [43]. Challenges persist in fundamental research, original innovation, and pivotal core technologies, with notable constraints in key components and major equipment procurement, leading to vulnerabilities in the industrial supply chain. Innovation remains a focal point, yet challenges persist in fundamental research, original innovation, and the cultivation of pivotal core technologies. These obstacles, coupled with constraints in procuring key components and major equipment, contribute to vulnerabilities in Shenzhen's industrial supply chain, impacting its resilience and adaptability in a globalized market [3]. Ecological and livelihood aspects exhibit shortcomings, with education, healthcare, and housing inadequately meeting the demands of rapid population expansion [57]. The urban landscape, production methods, organizational structures, and operational frameworks have undergone significant transformations, presenting numerous challenges such as heightened urban governance pressures and limited development space [43], rendering the task of integrating development and security complex and formidable.

Shenzhen's economic and social advancement is encountering various hurdles and complexities. Firstly, the external environment remains intricate and challenging, characterized by insufficient effective demand and an uncertain economic recovery foundation [8], necessitating further bolstering of market confidence. Secondly, attention must be directed towards fortifying the independence and control of key core technologies, while enhancing the resilience and modernisation of industry chain supply chains [32]. Thirdly, there's a need to reinforce the forward-looking and integrative role of urban land space planning, alongside improving the efficiency and intensive utilisation of land resources. Fourthly, there's a call to enhance the provision of high-quality public services, bridging the gap between public expectations and the desired standards in medical and healthcare, childcare services, and housing security. Fifthly, there's an imperative to elevate the scientific, refined, and intelligent governance of mega-cities, enhancing the efficacy of government services and administration. Confronting these challenges directly demands resolute measures and proactive approaches to address them effectively. Confronting these challenges head-on requires a concerted effort to implement stronger measures. It necessitates proactive strategies to foster innovation and resilience in the face of external uncertainties. Additionally, collaborative initiatives are imperative to enhance public services, bridge societal gaps, and elevate governance standards, ensuring Shenzhen's continued progress and prosperity amidst evolving economic and social landscapes.

6 Key Lessons Learnt from Shenzhen for Other Cities/contexts

Shenzhen stands as a beacon of urban development, offering key lessons that transcend its boundaries and resonate with cities worldwide. Undoubtedly, one of the paramount lessons is the power of embracing innovation and experimentation. The city's significance of strategic planning and infrastructure investment means extended development of its relatively new urban landscape and robust transportation networks that exemplify the dividends of foresight and long-term investment in infrastructure development. Also, Shenzhen's success highlights the value of fostering a conducive business environment. Its commitment to market-oriented reforms, streamlined regulatory processes, and incentives for entrepreneurship has attracted businesses and talent from around the world, driving its economic ascendance. As expressed earlier, Shenzhen exemplifies the power of collaboration and openness. Its strategic location in the Greater Bay Area and close ties with neighbouring Hong Kong SAR highlight the benefits of regional cooperation and integration in driving economic growth and innovation. In sum, we would like to summarise a few key lessons learnt for other cities in China and in other contexts/regions/countries. It is just fair to say that Shenzhen provides valuable lessons for other cities worldwide, and a few that we highlight here.

As the book title entails, recent urban changes are of our interest to see what we could learn from Chinese cities. The transition they have gone through could be a great opportunity of learning for developing nations of the global south. Therefore, Shenzhen's rapid urbanization and economic diversification are worth highlighting as primary lessons learnt. Shenzhen's swift transition from a fishing village to a global tech hub emphasises the benefits of embracing rapid urbanisation and diversifying the economy. Shenzhen's story of recent three to four decades show cities can learn to adapt quickly to changing economic landscapes. The other key lesson associated to Shenzhen's rapid development is its strategic plan to become a Special Economic Zone (SEZ) [58]. Creating specialized economic zones, akin to Shenzhen's pioneering model, has the potential to allure foreign investment, foster innovation, and ignite economic expansion [24]. This approach is not only effective but also adaptable for cities seeking to catalyse their economic progress. By designating specific areas for targeted development and offering incentives such as tax breaks and streamlined regulations, cities can create environments conducive to business growth and innovation. Furthermore, these zones serve as incubators for experimentation and collaboration, facilitating the exchange of ideas and technologies that drive economic advancement. Thus, the establishment of special economic zones represents a versatile strategy that empowers cities to leverage their unique strengths and aspirations to propel sustainable development and prosperity.

As highlighted throughout the chapter, Shenzhen's success stems from its emphasis on innovation and technology. This serves as a valuable lesson for other urban centers/environments/regions/hubs, highlighting the critical significance of prioritising investments in research and development. By cultivating an environment that nurtures entrepreneurship and fosters technological progress, cities can follow Shenzhen's trajectory towards prosperity. Encouraging collaboration between academia, industry, and government entities becomes imperative, fostering a culture that values innovation and embraces cutting-edge advancements. Additionally, investing in education and workforce development initiatives ensures a steady pipeline of skilled talent capable of driving technological innovation forward. Through these concerted efforts, cities can position themselves as dynamic hubs of innovation, driving economic growth, and enhancing global competitiveness in the rapidly evolving digital age. Thus, Shenzhen's success story serves as a beacon of inspiration, guiding other cities towards a future marked by innovation-driven prosperity and sustainable development.

Another crucial aspect to consider is the equilibrium between economic expansion and sustainability. Shenzhen's dedication to sustainable urban planning exemplifies the significance of harmonising economic growth with environmental preservation [17]. Their strategic plans after reaching a high level of development started to focus more on environmental qualities. Other urban centers can draw valuable lessons from this approach to cultivate liveable and environmentally conscious communities. Notably, successful instances of this approach are often realized through robust Public–Private Partnerships (PPPs). Shenzhen's developmental progress has been characterised by productive collaborations between governmental entities and private enterprises. This collaborative framework stands as a shining example of how partnerships can catalyse urban development and strengthen economic advancement [33]. By leveraging the strengths of both sectors, cities can establish innovative solutions that prioritise environmental stewardship while fostering economic prosperity [12]. Hence, Shenzhen's experience offers a blueprint for nurturing synergistic partnerships that drive sustainable urban growth, setting a precedent for cities worldwide to learn from. This aligns well with the promotion of global integration and connectivity agenda. Shenzhen's closeness to Hong Kong has facilitated global connectivity for the city and its region. Other cities can learn the advantages of strategic geographical positioning and integration with neighboring regions for enhanced economic and cultural exchange.

Furthermore, Shenzhen has been successful in developing and sustaining flexible and adaptive governance structure, leading to a more optimised operations between different units, too [13]. Shenzhen's success is attributed, in part, to its adaptability and flexibility in governance, allowing for quick responses to changing economic and technological landscapes. This agility is crucial for cities facing dynamic challenges. In addition, Shenzhen has invested extensively in human capital, particularly younger and more educated workforces whom have been attracted by the cluster of industries in the vicinity of Shenzhen. Shenzhen's emphasis on education and talent development contributes to its innovation ecosystem. Other cities can learn to invest in human capital to create a skilled workforce capable of driving technological advancements. Lastly, we can summarise that by studying Shenzhen's urban change and development over these few decades, cities worldwide can collect insights into strategies for sustainable urban development, economic growth, and fostering innovation in a rapidly changing global landscape.

References

1. Bontje M (2014). Creative Shenzhen? A critical view on Shenzhen's transformation from a low-cost manufacturing hub to a creative megacity. Int J Cult Creat Ind 1(2): 52–67
2. Bulanda-Jansen A (2019, February) Evolution of the metropolitan area of Shenzhen, analysis: from theory to selected examples. In: IOP conference series: materials science and engineering, vol 471, no 11. IOP Publishing, p 112058
3. Caprotti F, Cowley R, Bailey I, Joss S, Sengers F, Raven R, Spaeth P, Jolivet E, Tan-Mullins M, Cheshmehzangi A, Xie L (2017) Smart eco-city development in Europe and China: opportunities, drivers and challenges. https://ore.exeter.ac.uk/repository/handle/10871/29876
4. Chang Z, Diao M (2022) Inter-city transport infrastructure and intra-city housing markets: estimating the redistribution effect of high-speed rail in Shenzhen, China. Urban Stud 59(4):870–889.
5. Cheshmehzangi A (2020) Evaluating the nexus between housing and energy sectors: the comparison of urban, Peri-urban and rural housing areas in Zhuhai, China. Energy Power Eng 12(06):314–338.
6. Cheshmehzangi A (2020) Identity of cities and city of identities: examples. In Identity of cities and city of identities, Springer, Singapore, pp 179–244
7. Cheshmehzangi A, Chen H (2021) Low carbon transitions: practices and lessons. In China's sustainability transitions: low carbon and climate-resilient plan for carbon neutral 2060. Springer, Singapore, pp 119–137
8. Cheshmehzangi A, Dawodu A, Sharifi A (2021) Sustainable urbanism in China. Routledge, New York
9. Cheshmehzangi A, Tang T (2022) Pearl river delta city cluster: from dual-core structure economic development strategies to regional economic plans. In: China's city cluster development in the race to carbon neutrality. Springer Nature, Singapore, pp 63–75
10. Cheshmehzangi A, Tang T (2022) China's city cluster development: from megacities development trend to urban agglomeration plan. In: China's city cluster development in the race to carbon neutrality. Springer Nature, Singapore, pp 25–34
11. Cheshmehzangi A, Tang T (2022) China's city cluster development: from megacities development trend to urban agglomeration plan. In: China's city cluster development in the race to carbon neutrality. Springer Nature, Singapore, pp 25–34
12. Cheshmehzangi A, Tang T (2022) Pearl river delta city cluster: from dual-core structure economic development strategies to regional economic plans. In: China's city cluster development in the race to carbon neutrality. Springer Nature, Singapore, pp 63–75
13. Cheshmehzangi A, Tang T (2022) Guangdong-Fujian-Zhejiang coastal region: a network corridor between three coastal provinces. In: China's city cluster development in the race to carbon neutrality. Springer Nature, Singapore, pp 119–133
14. Cheshmehzangi A, Xie L, Tan-Mullins M (2018) The role of international actors in low-carbon transitions of Shenzhen's international low carbon city in China. Cities 74:64–74
15. Cirera X, Mason AD, De Nicola F, Kuriakose S, Tran TT (2021) The innovation imperative for developing East Asia. World Bank Publications, Washington DC
16. Deng W, Cheshmehzangi A (2018) Macro level: eco-city cases in China. In: Eco-development in China: cities, communities and buildings. Palgrave Macmillan, Singapore and Germany, 105–155
17. Deng Z (2009) Developing urban design as public policy in China: a case study in Shenzhen. Cardiff University, United Kingdom. https://search.proquest.com/openview/189c626db93897684e5f6670790ce119/1?pq-origsite=gscholar&cbl=2026366

18. Du J (2019) Demystifying Shenzhen: the informal in a planned city. Doctoral dissertation, ETH Zurich
19. Du J (2020) The Shenzhen experiment: the story of China's instant city. Harvard University Press, Cambridge
20. Hai J (2023) Flourish an innovation tomorrowland: the local developmental state model and China's high-tech park. https://kclpure.kcl.ac.uk/portal/files/250583269/2024_Hai_Jiawei_1413273_ethesis.pdf
21. Hao L (2023) The impact of government support to enterprises in emerging industries: the case of Shenzhen. https://repositorio.iscte-iul.pt/handle/10071/28534
22. Hu A (2023) The great leap development and outlook of China's scientific and technological strength (2000–2035). Front Bus Res China 17(3)
23. Hu R (2019) The state of smart cities in China: the case of Shenzhen. Energies 12(22):4375
24. Kwok RYW, Chan TM (1991) Economic development in the. Shenzhen special economic zone: appendage To Hong Kong? Asian J Soc Sci 19(1):180–205
25. Li H, Chen P, Grant R (2021) Built environment, special economic zone, and housing prices in Shenzhen, China. Appl Geogr 129:102429
26. Li J (2024) Impact of financial development on innovation efficiency of high-tech industrial development zones in Chinese cities. Technol Soc 76:102467
27. Li Y (2021) New actors in the old hierarchies: alliances for low-carbon urban development in Shenzhen, China. J Plan Educ Res. 0739456X211002894
28. Lin Z (2023) The changing Chinese intellectual property landscape in the new era: understanding the connections between corporate IP practices and business performance in Shenzhen. https://repositorio.iscte-iul.pt/handle/10071/28536
29. Liu C (2013) Gown and town: university spatial development and urban transformation in the knowledge society. The case of Shanghai. Politecnio Milan, Milan https://www.politesi.polimi.it/handle/10589/75104
30. Liu C, Cai Y (2018) Triple Helix model and institutional logics in Shenzhen special economic zone. Sci Public Policy 45(2):221–231
31. Liu H (2005) China's special economic zones: from Shenzhen to Shanghai. A new path to industrialization, urbanization, globalization and modernization. New School University. https://search.proquest.com/openview/ab0a97b36e9d4efdc87e44ff4ffd9ac7/1?pq-origsite=gscholar&cbl=18750&diss=y
32. Ma S, Ding J, Huang Z, Guo R (2024) Evaluation of the urban industrial coupling strategy based on the global production networks theory: a case study of the smart phone industry in the Guangdong-Hong Kong-Macao Greater Bay Area. PLoS ONE 19(4):e0300588
33. Mangi E, Chen W, Heath T, Cheshmehzangi A (2024) Prioritizing sustainable urban regeneration practices: addressing contemporary urban challenges through the lens of public participation. In: innovative public participation practices for sustainable urban regeneration. Springer Nature, Singapore, pp 1–11
34. NetEase Finance (2023) Analysis: Shenzhen's top ten advantages and its role as a "button" in the Greater Bay Area. <<分析:深圳十大优势及其在大湾区的"纽扣"作用>>. https://q.stock.sohu.com/cn/gg/2024/688589/57172272.shtml
35. Ng MK (2011) Strategic planning of China's first special economic zone: Shenzhen city master plan (2010–2020). Plan Theory Pract 12(4). https://www.tandfonline.com/doi/full/10.1080/14649357.2011.626316
36. Ng MK, Tang WS (2004) Theorising urban planning in a transitional economy: the case of Shenzhen, people's republic of China. Town Plan Rev 173–203
37. Ng MK, Wong C (2022) Spatial planning for smart sustainable development? Plan Theory Pract 23(5):759–798
38. O'Donnell MA (1999) Path breaking: constructing gendered nationalism in the Shenzhen special economic zone. Positions: East Asia Cult Crit 7(2):343–375
39. Pan W, Du J (2021) Towards sustainable urban transition: a critical review of strategies and policies of urban village renewal in Shenzhen, China. Land Use Policy 111:105744

40. Phillips DR, Yeh AGO (1983) China experiments with modernisation: the Shenzhen special economic zone. Geography 289–300
41. Reil S, Hooks C (2020) Understanding the Shenzhen "Miracle": a field study of urban and economic development in the Shenzhen SEZ. J Spec Jurisd 1(1):27–27
42. Shen J, Kee G, Shen J, Kee G (2017) Shenzhen: innovation and governments' roles in reform and development. In: Development and planning in seven major coastal cities in southern and eastern China, pp 61–92.
43. Shenzhen Municipal People's Government (2021) Shenzhen municipality's 14th five-year plan for national economic and social development and outline of long-term goals for 2035. <<深圳市国民经济和社会发展第十四个五年规划和二〇三五年远景目标纲要>>. http://www.sz.gov.cn/cn/xxgk/zfxxgj/ghjh/content/post_8854038.html
44. Shenzhen Municipal People's Government (2023) 2024 Government work report. <<2024年政府工作报告>>. http://www.sz.gov.cn/gkmlpt/content/11/11142/post_11142744.html
45. So AY (1988) Shenzhen special economic zone: China's struggle for independent development. Can J Dev Stud /revue Canadienne D'études Du Développement 9(2):313–323
46. Sugrue TJ (2017) Making cities global: the transnational turn in urban History. University of Pennsylvania Press, Pennsylvania
47. Sun H (1991) Urban development in Shenzhen SEZ. Habitat Int 15(3):25–31
48. Tantri ML (2013) Trajectories of China's integration with the world economy through SEZs: a study of Shenzhen SEZ. China Report 49(2):227–250
49. Tao Y, Lu Z, Tao Y, Lu Z (2018) Cultural contributions from the special economic zones to China's development path. In: Special economic zones and china's development path, pp 233–274
50. Wang Y (2022) Exploring knowledge exchange, innovation and collaboration in creative industries: a comparative analysis between Shenzhen and Brisbane through the triple helix framework. Doctoral dissertation, Queensland University of Technology
51. Wei Q, Zhang Y (2024) Vertical expansion in the making: planning against deindustrialization by promoting "Industry's Going Upstairs" in Shenzhen. Environ Plan A: Econ Space. 0308518X241226889.
52. Wu LF (2020) Evaluating China's efforts to foster innovation in the 21st century. Doctoral dissertation, Naval Postgraduate School
53. Wu W (2005) Dynamic cities and creative clusters, vol 3509. World Bank Publications, Washington DC
54. Wu W (2018) Pioneering economic reform in China's special economic zones: the promotion of foreign investment and technology transfer in Shenzhen. Routledge, New York
55. Xie L, Cheshmehzangi A, Shao M, Zhang Y, Chan F (2023) Urban sustainability experiments in China: plural approaches for transformation. In: Handbook on China's urban environmental governance. Edward Elgar Publishing, Cheltenham, pp 169–185
56. Xie L, Cheshmehzangi A, Tan-Mullins M, Flynn A, Heath T (2020) Urban entrepreneurialism and sustainable development: a comparative analysis of Chinese eco-developments. J Urban Technol 27(1):3–26
57. Xie L, Tan-Mullins M, Cheshmehzangi A (2019) Political ecology of Chinese smart eco-cities. In: Remaking sustainable urbanism: space, scale and governance in the new urban era, pp 57–78
58. Yee FLW (1992) Economic and urban changes in the Shenzhen special economic zone, 1979–1986. Doctoral dissertation, University of British Columbia
59. Yejoo K (2013) Chinese-led SEZs in Africa are they a driving force of China's soft power. Discussion Paper. https://scholar.sun.ac.za/handle/10019.1/85183
60. Yeung H (2015) A tale of two cities–the development and reform experiences of Shenzhen and Shanghai. J Chin Econ Bus Stud 13(4):369–396
61. Zacharias J, Tang Y (2010) Restructuring and repositioning Shenzhen, China's new mega city. Prog Plan 73(4):209–249
62. Zhang LE, Zhao S, Kern P, Edwards T, Zhang ZX (2023) The pursuit of indigenous innovation amid the tech cold war: the case of a Chinese high-tech firm. Int Bus Rev 32(6):102079

Chapter 5
Guangzhou: China's Strategic Southern Gate with a Global Outlook for Innovation

1 General Information and Introduction to Guangzhou

Guangzhou, situated in the southern part of China, boasts a captivating history that spans more than 2200 years, tracing its origins back to the Qin Dynasty [33]. This historical significance has shaped the city's identity and contributed to its multifaceted character. As a key port for international trade, Guangzhou has been a hub of commerce, attracting merchants from distant corners of the world [34, 50]. Its cosmopolitan nature reflects the exchange of goods, cultures, and ideas that have occurred within its lively streets. From ancient times, Guangzhou's strategic location along the Pearl River Delta has facilitated trade routes and positioned it as a gateway to the global marketplace [40, 61]. Today, as the capital of Guangdong Province, Guangzhou plays a crucial role in China's economic landscape. Its bustling markets, iconic landmarks, and dynamic atmosphere make it a vibrant center for business, culture, and innovation [11, 30]. The city's proximity to Hong Kong and Macau (Figs. 1–3) further enhances its economic significance and connectivity.

This blend of history, economic prominence, and cultural richness positions Guangzhou as a multifaceted and influential city in the region [18, 23, 54]. The city's strategic location along the Pearl River Delta has contributed to its status as a major trading hub and a gateway to the rest of the world [5]. The city's strategic location along the Pearl River Delta has further augmented its status as a leading trading hub and a gateway connecting China to the global marketplace, contributing to its reputation as a dynamic and cosmopolitan metropolis [16]. Its proximity to Hong Kong and Macau further enhances its economic significance. With a population exceeding 14 million people, Guangzhou is a melting pot of cultures and influences [31]. It attracts visitors from around the globe who come to explore its diverse culinary scene, vibrant festivals, and dynamic arts and entertainment offerings. Overall, Guangzhou's blend of history, economic prominence, and cultural richness makes it a fascinating destination.

A. Cheshmehzangi and T. Tang, *30 Years of Urban Change in China's 10 Core Cities*, Urban Sustainability, https://doi.org/10.1007/978-981-97-8846-0_5

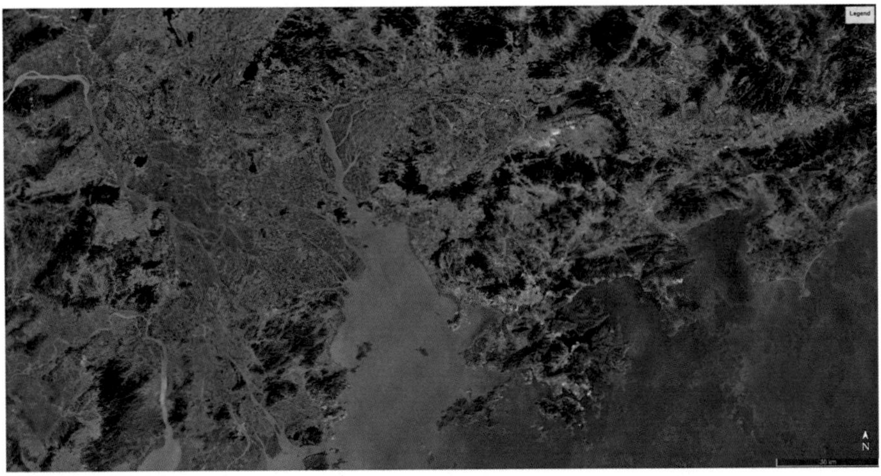

Fig. 1 Guangzhou and strategic coordinating positioning in the Greaer Pearl River Delta, alongside Macao and Hong Kong SARs. Development plans for recent three decades and changes in this region indicate it significance. Map is from 1990s to 2000s. (Maps are extracted from open and accessible maps on Google Earth, generated in April 2024)

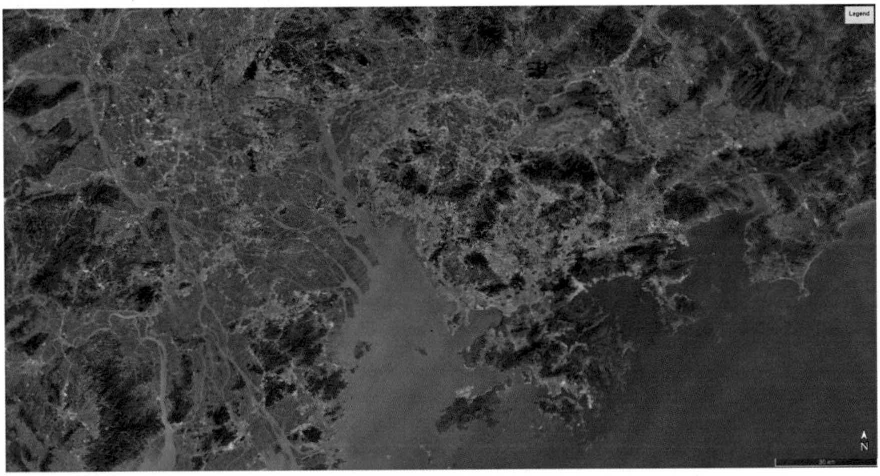

Fig. 2 Guangzhou and strategic coordinating positioning in the Greaer Pearl River Delta, alongside Macao and Hong Kong SARs. Development plans for recent three decades and changes in this region indicate it significance. Map is from 2010s. (Maps are extracted from open and accessible maps on Google Earth, generated in April 2024)

Fig. 3 Guangzhou and strategic coordinating positioning in the Greaer Pearl River Delta, alongside Macao and Hong Kong SARs. Development plans for recent three decades and changes in this region indicate it significance. Map is from early 2020s. (Maps are extracted from open and accessible maps on Google Earth, generated in April 2024)

Guangzhou's development from ancient times to its current status as a dynamic metropolis highlights its resilience and adaptability [52]. Over the centuries, the city has endured the fall and flow of many dynasties, navigated through periods of conflict and disruption, and emerged as a symbol of progress and prosperity in southern China [8]. Its strategic significance as a trading nexus has only intensified over time, cementing its role as a pivotal economic conduit (see Figs. 1–4) linking China with the global community [19]. The architectural panorama of Guangzhou serves as evidence to its rich historical and cultural legacy. Here, ancient temples coexist harmoniously with modern skyscrapers, symbolising the city's enduring development through time. Today, Guangzhou stands as one of China's premier urban centers, flourishing as a nexus of commerce, culture, and innovation [24]. Its vibrant streets and bustling markets beckon both visitors and entrepreneurs alike, embodying the city's dynamic spirit.

1.1 Guangzhou's Physical and Main Development Over the Past 30 years

In recent three decades, Guangzhou's economic growth has been steady and optimised, showcasing China's main economic center in the south. The gross regional product has exceeded 2.5 trillion yuan, with an average annual growth rate of 6%, and the per capita gross regional product has exceeded 20,000 yuan, reaching the level of middle-developed countries [47, 59]. The industrial structure has been continuously

Fig. 4 Guangzhou in 1985 was still a major regional city with strong logistics and connectivity nationwide and internationally (Map is extracted from open and accessible maps on Google Earth, generated in April 2024)

optimised [17], and a number of major projects such as high-definition displays, integrated circuits [51], intelligent and new energy vehicles, and biomedicine and health have been successfully introduced and put into operation [9, 13]. The output value of high-tech products above the scale accounts for 50 percent of the total output value of industrial products above the scale. In these three decades, and perhaps more in recent years, new technologies, new industries, and new business forms/patterns have flourished across the city [26]. This pattern of development has made Guangzhou a very diverse metropolis where each district has developed its own characteristics based on the city's capacity of trade and business [20]. The balance of local and foreign currency deposits and loans of financial institutions exceeded 12 trillion yuan, and the Guangzhou Futures Exchange was approved for establishment. Such advancements are unique in China, indicating the city has already gained its national and international status (see Figs. 5–8).

For Guangzhou, these three decades of development signify progressive development, meaning that innovation and development have achieved remarkable results. The concept of innovation has been integrated into all fields and processes of economic and social development. Layout construction of "one district and three cities" has paved a path for smooth preparation of the construction of national laboratories in the field of respiratory diseases [26]. Also, four major scientific and technological infrastructures such as human cell lineage were completed through the preliminary work of pre-research. In addition, four provincial laboratories such as Bio-Island have been successfully secured to build a whole-chain innovation and development pathway. In only a few years, the number of high-tech enterprises increased from 1919 to 12,000, and the number of national science and technology-based small and medium-sized enterprises (SMEs) filed into the database for the past three years has accumulated to be the first in the country [26]. The increase in the

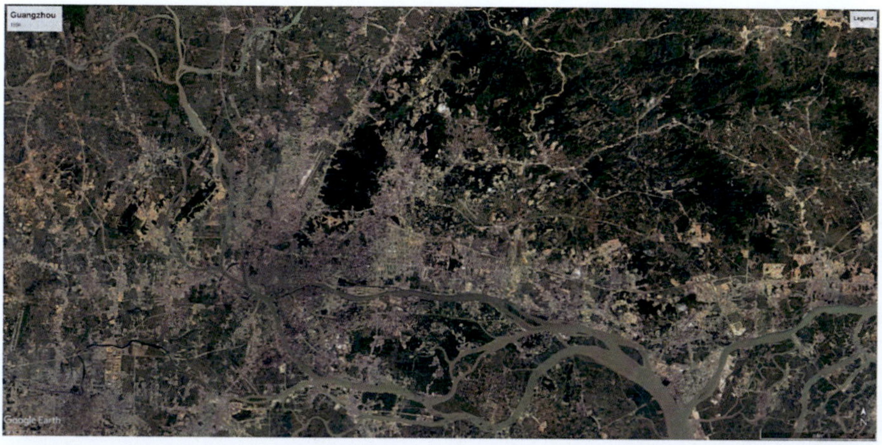

Fig. 5 Guangzhous expansion and development over three decades showing its vast scale of development, densification and in-fill development, and infrastructural upgrades. Map is from 1990s. (Maps are extracted from open and accessible maps on Google Earth, generated in April 2024)

Fig. 6 Guangzhous expansion and development over three decades showing its vast scale of development, densification and in-fill development, and infrastructural upgrades. Map is from early 2000s. (Maps are extracted from open and accessible maps on Google Earth, generated in April 2024)

intensity of R&D investment by the whole society ranked first among major cities in China. This means there are excellent investment and opportunities that put the Guangzhou-Shenzhen-Hong Kong Innovation Cluster ranked second in the world [26]. This is based on the fact that Guangzhou's innovation capacity is now ranked second in China among national innovative cities.

Furthermore, Guangzhou's energy level has jumped significantly in recent years. The city took the lead in the country in launching a pilot municipal spatial master

Fig. 7 Guangzhous expansion and development over three decades showing its vast scale of development, densification and in-fill development, and infrastructural upgrades. Map is from 2010s. (Maps are extracted from open and accessible maps on Google Earth, generated in April 2024)

Fig. 8 Guangzhous expansion and development over three decades showing its vast scale of development, densification and in-fill development, and infrastructural upgrades. Map is from 2024. (Maps are extracted from open and accessible maps on Google Earth, generated in April 2024)

plan, adding 165 square kilometers of new construction land and revitalising 67 square kilometers of stock of land for urban renewal [26]. The second terminal of Baiyun International Airport and the third phase of Nansha Port Area have been completed and opened, and the third phase of airport expansion and the fourth phase of Nansha Port Area have been accelerated. Nowadays, there are claims that Baiyun International Airport ranks first in the world in terms of passenger throughput. The port's cargo and container throughput rank fourth and fifth respectively in the world, while the volume of express delivery business ranks first among major cities in China. In recent years, and within less than a decade of development, the Guangzhou-Shenzhen-Hong Kong High Speed Railway, Guangshi Railway, Guangzhou-Qingdao Intercity Phase one and other projects have been completed and opened [26]. It is important to note that the volume of railroad passenger traffic is the second largest in China [10], meaning there are also regional importance of Guangzhou due to its national and international connectivity. The mileage of subway operation reached 531 km, ranking third in China.

Approved as a national pilot zone for the innovative development of new-generation artificial intelligence, Guangzhou has extended its investment in digital technologies and data-driven centers. It has emerged as a major regional hub focusing on both high-tech and digital industries. This investment and new attention have prompted the city to accelerate the construction of key functional zones such as Guangzhou Artificial Intelligence and Digital Economy Pilot Zone, Sino-Singapore Guangzhou Knowledge City, and Airport Economic Zone [26]. The performance appraisal of rural revitalisation strategy won the first place in the Pearl River Delta region for two consecutive years, and the Guangzhou-Qingdao Joint Area was listed as the first batch of national experimental zones for integrated urban and rural development.

In recent years, Guangzhou has progressed to deepen reform and opening up strategies at the city and regional levels. The latter level is critical since Guangzhou plays a pivotal role at the provincial level. The city was honoured as the 2020 National Benchmark City for Business Environment, with 2.697 million market entities, an increase of 1.03 times compared with 2015 [26]. The city officials have confirmed they fully promoted the construction of the Guangdong-Hong Kong-Macao Greater Bay Area [19], supported Shenzhen in building an advanced demonstration zone of socialism with Chinese characteristics, and further enhanced the function of the core engine of the "One Core, One Belt, One Region" [26]. For instance, Nansha Pilot Free Trade Zone system innovation achievements are counted as 43 in the country with 112 replicated innovation accomplishments in the province.

Strategically, Guangzhou actively participates in the construction of the "Belt and Road Initiative (BRI)", connecting other national cities with southern regions, Specially Administrative Regions (SARs) [38], and other countries in Asia and beyond [3, 22]. The total value of merchandise imports and exports exceeded 1 trillion yuan in 2019, representing a super hub for trade and business nationwide and internationally. Aside from the organisation of major international events, such as "Read China" International Conference (Guangzhou), "Fortune" Global Forum, and other international conferences, Guangzhou has shown a continuous track record of investment in the world's top 500 enterprises [26]. The total number has now reached 309 enterprises, leading to a wider recognition of the world's top 500 enterprises in Guangzhou [44]. As it stands, Guangzhou's ranking in the world's city system has historically crossed into the first rank. The latest report in November 2023 suggests the total revenue of the top 500 companies has already reached 5.6 trillion yuan [44], indicating their significant contribution to Guangzhou's overall revenue growth.

As it appears from the records, Guangzhou has been making significant strides in various aspects of governance and development. The city has effectively managed financial and government debt risks demonstrates prudent fiscal management, while successfully lifting all poor villages and individuals out of poverty showcases a strong commitment to social welfare and economic development. It also has won recognition like the "China's Special Contribution to Comprehensive Well-being" award for two consecutive years underscores the city's dedication to comprehensive progress and improvement [26]. These achievements speak volumes about Guangzhou's leadership and the collective efforts of its people. Guangzhou also has been made key progress in pollution prevention and control, with the proportion of good air quality days reaching 90.4%, six indicators meeting the standards for the first time, and the concentration of respirable particulate matter (PM2.5) dropping to 23 μg per cubic meter. This progression is optimal for the national central city, and the first batch of national black-smell water body treatment demonstration cities, with the water quality of the 13 national and provincial examination sections all meeting the standards [26]. The city has progressed extensively in managing domestic garbage classification to achieve full coverage of urban and rural areas. It is now selected as the first World Bank sustainable urban cooling pilot [26]. Based on the records, the city's energy saving and carbon reduction and the emission of major pollutants have completed the provincial target.

2 Guangzhou's Development Strategies and Directions

In its ongoing evolution, Guangzhou determinedly advances the development of the Guangdong-Hong Kong-Macao Greater Bay Area to a new level. This effort reinforces its role as a key driver and catalyst for high-quality development [43]. With a vision of national greatness, the city aims to seize the significant opportunities presented by the Greater Bay Area construction [21]. It seeks to enhance strategic connectivity, effectively broadening economic integration within the region and beyond. By leveraging national strategies, Guangzhou plans to translate potential into dynamic momentum for top-tier development [25]. The city also adheres to the industry as its priority, ensuring that it remains as a crucial regional manufacturing city with substantial planting economic development advantages [6, 46]. The city maintains a commitment to prioritising economic development in the real sector, emphasising the transformation and modernisation of conventional industries, expediting the growth of emerging sectors, and swiftly strategising for future industrial development and distribution [1, 2]. These efforts aim to hasten the establishment of innovative and robust productivity frameworks, laying a solid foundation for stable and high-quality development.

It is evident that initiatives have been undertaken to augment Guangzhou's domestic demand and secure external demand stability, aiming to strengthen its inherent drivers of economic growth. Key stakeholders and governmental entities within the city prioritize reinforcing growth stability, actively implementing strategies to expand domestic demand, fostering synchronised development of both internal and external demand, and supplementing the supportive and leading role of seamless domestic and international economic circulation. Such direction has already helped strengthening the support of science and technology, education and talents, and shape the new momentum and new potential for development [25]. Therefore, the city has pledged to adhere to science and technology as its first productive force, talent as the first resource, and innovation as its first driving force. In doing so, it aims to push forward the construction of a strong city of science and technology innovation [25], education and talent as a whole, so as to build an important bearing place of international science and technology innovation center.

Furthermore, Guangzhou pledges to emphasise its leadership in ecological construction within the framework of "Green and Beautiful Guangzhou," aspiring to fashion a picturesque and hospitable "Flower City" [25]. Government officials have committed unwaveringly to pursuing a civilized developmental trajectory characterised by productive growth, prosperous livelihoods, and ecological harmony. They pledge to synergistically advance initiatives for carbon and pollution reduction, as well as green expansion and prosperity, with the aim of crafting a Guangzhou characterised by green landscapes, blue skies, and pristine waters [25]. Through these concerted efforts, the city seeks to enhance its environmental sustainability [48] while simultaneously fostering a high quality of life for its residents (see Figs. 9–11), in line with its vision for a harmonious and prosperous future.

Fig. 9 An example of intensive development and land use change in one of Guangzhou's central areas—from 2000 to 2023. Map is from 2000. (Maps are extracted from open and accessible maps on Google Earth, generated in April 2024)

Fig. 10 An example of intensive development and land use change in one of Guangzhou's central areas—from 2000 to 2023. Map is from early 2010s. (Maps are extracted from open and accessible maps on Google Earth, generated in April 2024)

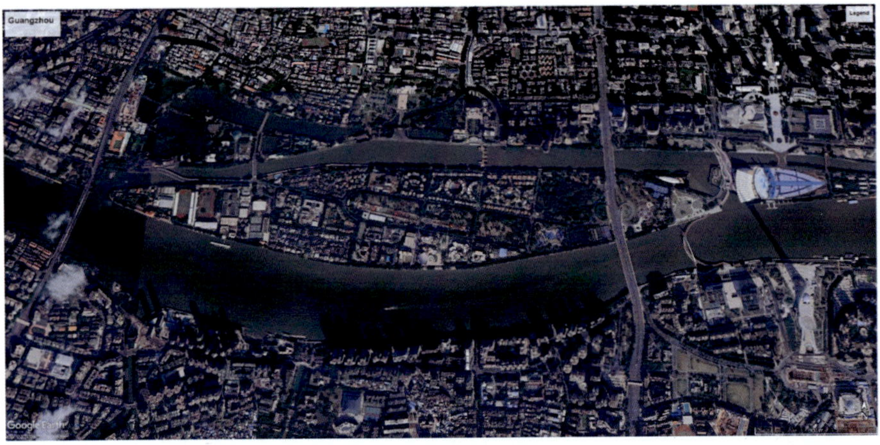

Fig. 11 An example of intensive development and land use change in one of Guangzhou's central areas—from 2000 to 2023. Map is from 2023. (Maps are extracted from open and accessible maps on Google Earth, generated in April 2024)

2.1 Development Plan for Guangzhou Transport Project

In 2018, the realisation of the "district metro" in Guangzhou marked a significant milestone with the completion and opening of Line 14's initial phase [49]. Over the subsequent five years leading up to 2022, a remarkable expansion unfolded with the introduction of several new lines and extensions, notably Line 21, the Yan Gang to Lijiao section of Guangfo Line, the northern stretch of Line 8, inaugural segments of

Line 18 and Line 22, and the westward extension of Line 7 [49]. These developments substantially bolstered the city's metro network, increasing its total length from 390.6 km to 621 km, greatly enhancing transportation accessibility for the residents of Guangzhou.

Over the course of these five years, Guangzhou Metro has not only achieved significant advancements in urban rail transit systems like the metro and tram but has also made notable strides in comprehensive urban rail transit, including intercity railroads [27, 57]. On June 11, 2019, the establishment of the wholly owned subsidiary, Guangdong Intercity Operation Company Limited, signified a considerable leap for Guangzhou Metro towards the exploration of integrated operations encompassing "Metro + Intercity" [49]. This move was a pivotal advancement in public transportation infrastructure of the city. Then, on November 30, 2020, Guangzhou Metro Group achieved another milestone with the inauguration of two intercity lines: Guangzhou-Qingdao Intercity (from Huadu Station to Qingcheng Station) and Guangzhou Donghuan Intercity (from Huadu Station to Baiyun Airport North Station) (see Figs. 12 and 13) [49]. This accomplishment positioned Guangzhou Metro Group as the first local enterprise in China to independently operate intercity railroads [55], marking a momentous achievement in its journey of transit expansion and integration.

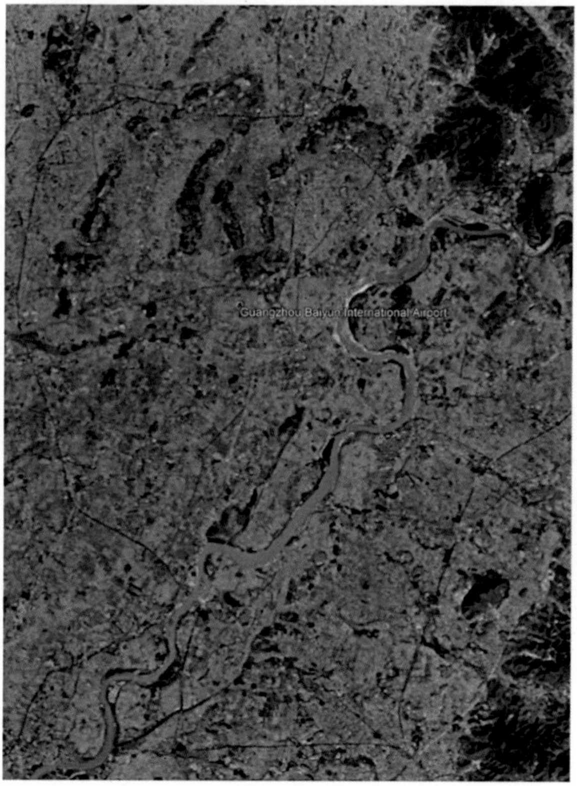

Fig. 12 An example of large scale development for transportation links, such as Guangzhou Baiyun International Airport in North Guangzhou. Map is from 1980s. (Maps are extracted from open and accessible maps on Google Earth, generated in April 2024)

Fig. 13 An example of large scale development for transportation links, such as Guangzhou Baiyun International Airport in North Guangzhou. Map is from early 2020s. (Maps are extracted from open and accessible maps on Google Earth, generated in April 2024)

In addition to the operation business, Guangzhou Metro has achieved a significant breakthrough in the construction sector, particularly concerning intercity railroads. By 2022, adhering to provincial and municipal directives, Guangzhou Metro actively directed inter-city rail projects within the Pearl River Delta (PRD) region and assumed control of the PRD Intercity Railway Company [49]. This successful takeover and subsequent project developments serve to leverage Guangzhou Metro Group's specialised capabilities in rail investment, construction, development, and management. This initiative aims to establish a comprehensive rail transportation network in the Guangdong, Hong Kong, and Macao Greater Bay Area, fostering seamless connectivity under the "*one network, one ticket, one string of cities*" model [49]. These efforts align with the vision of creating a well-connected rail infrastructure in the Greater Bay Area [36], thereby facilitating efficient travel, and supporting overarching objectives such as the integration of transportation networks and achieving a 60-min direct rail connection [32] between Guangzhou and other cities in the Bay Area.

Currently, Guangzhou Metro Group oversees the operation of a substantial rail transit network spanning 776 km. This encompasses not only the local metro network covering 621 km, inclusive of the Guangzhou-Foshan line, but also tram routes totaling 22.1 km. Additionally, the group manages intercity railroads spanning 60.8 km [49]. Furthermore, it is involved in various international projects, such as the Pakistan Lahore Orange Line, Jiangxi Nanchang Metro Line 3, Hainan Sanya tram, Huawei Dongguan, Songshan Lake Kunming, and Changshui International Airport Passenger Express System, collectively stretching over 72.8 km. Moreover, the group is engaged in field projects like the Rapid Transit System (RTS), adding another 72.1 km to its portfolio [49]. This comprehensive network emphasises Guangzhou Metro Group's decisive role in advancing rail transportation [4] both domestically and internationally.

Simultaneously, Guangzhou Metro is actively advancing the development of 10 sections, totaling 201.9 km, within its metro line network. Additionally, it is spearheading the construction of nine sections, encompassing 316 km, for projects like the Foguan Intercity railway [49]. Moreover, the company bears the responsibility for the comprehensive investment and construction management of 24 national railway initiatives, comprehensive transportation hubs, and municipal road projects [49]. These endeavours highlight Guangzhou Metro's multifaceted role in shaping the region's transportation landscape, aiming not only to enhance intra-city connectivity but also to foster intercity mobility and facilitate seamless travel across various modes of transportation infrastructure [49]. Through these efforts, Guangzhou Metro continues to play a pivotal role in driving the sustainable development and integration of transportation networks in the region.

2.2 The Local Government Projection for Guangzhou's Future Development

Looking forward to 2035, the city aims to lead in achieving socialist modernization, with plans focused on bolstering economic prowess and advancing scientific and technological capacities within the city and its surroundings. These strategies also aim to enhance comprehensive competitiveness through future development blueprints. It is envisaged that Guangzhou will experience significant leaps in its gross regional product and per capita income for both urban and rural residents, elevating its per capita gross regional product to new heights [26]. The city aims to evolve into an international metropolis exuding classic charm and contemporary vitality [25]. This transformation will position Guangzhou as a global trade hub, a comprehensive transportation nexus, and a pivotal center for science, technology, education, culture, and healthcare, wielding substantial global influence. Embracing a central strategy of becoming a beautiful and liveable "Flower City" [28, 42, 45], Guangzhou aspires to emerge as a dynamic global hub, solidifying its place on the international stage.

As part of the city's blueprint for its further development, the modern industrial system will be more competitive [39]. In doing so, it aims to have major breakthroughs in key core technologies, and by achieving them, the city will be comprehensively built into a strong city with international competitiveness in science and technology innovation, advanced manufacturing industry, modern service industry, and talents, realising new-type industrialisation [18, 23], informatisation, urbanisation [14], and modernisation of agriculture [12]. It is noted that a number of headquarters and head enterprises will emerge to drive the development of innovation and to support the supply chain of the global industrial chain. Aligned with its strategic development plans, the city's hub function is aimed to be stronger [26]. This approach suggests strengthening the status of world-class airports, seaports and railroad hubs in a longer-term development process. In achieving these milestones, the city will become a leading international information hub, the gathering and radiation capacity will be stronger, and the digital transformation of economic and social development will be realised.

Furthermore, the city has proposed to be more internationalised through a higher degree of opening up to the outside world. This has already been developing in recent decades and is expected to form the regional economic agglomeration of southern China. In this regard, the status of the "southern gate" of China will be further consolidated and strengthened [26]. This approach will help establish and include the advantages of international cooperation and competition that are significantly enhanced through international strategies, marketisation plans, and global networks to the city.

A tangible direction for Guangzhou and making the city smarter is to ensure its urban governance is further modernised. The city proposes to fully guarantee the rights of the general public to equal participation and development. In doing so, the rule of law will be basically established in Guangzhou, and the construction of 'Safe Guangzhou' will reach a higher level [26]. The city culture will be more prosperous, and an international first-class city of culture, education, sports and health will be built. This direction is heavily politicised through socialist material civilization agenda and spiritual civilization that is coordinated as part of the city and regional progression. For instance, the degree of civilization of the city and the civilization of the citizens are planned to be significantly improved. There is a consolidated movement so that the comprehensive strength of the city's culture will be significantly enhanced. This approach enables the city to be an example for a strong socialist cultural power.

In addition, the overarching plan of 'Beautiful Guangzhou' is more attractive, suggesting a more holistic approach to have new pattern of harmonious coexistence of man and man, and man and nature [15]. This approach aligns well with national level agenda of ecological civilization that was announced in 2007 [26]. Over the years, and aligned with such strategies, green production and lifestyle have basically taken shape. Such direction suggests steadily decline of carbon emissions after reaching the proposed carbon peak in 2030. It is already evidenced by the local government and its agencies that the ecological environment has fundamentally changed for the better. There are ongoing work to form and reform the spatial pattern of land and space along the lines of high-quality development. The city officials have pledged that in coming years people's lives will be better, the gap between urban and rural development and the living standards of residents will be significantly narrowed, and the quality of 'Happy Guangzhou' [29] will be higher [26]. Altogether, such approach indicates the comprehensive development of human beings and the common prosperity of the people. With this at the core of its development agenda, Guangzhou plans to continue taking the lead in making more obvious and substantial progress.

3 Guangzhou's Challenges of Its Past and Present Progression

The primary argument asserts that Guangzhou's economic resilience remains insufficient, indicating a deficiency of high-level enterprises. Consequently, industries, particularly private enterprises, face mounting pressure for transformation and upgrading, resulting in production and operational challenges. Compared to larger global cities, there is a lack of robust expectations and untapped consumption potential. Secondly, there is a pressing need to strengthen innovation and achieve breakthroughs, which have been lacking for years, if not decades. It is observed that while there have been some innovative reforms, they have not been extensive enough to stimulate widespread societal innovation or foster strong entrepreneurial activity. Government reports highlight a suboptimal economic environment in Guangzhou over the years, indicating imperfect financial ecology, inadequate alignment between supply and demand of scientific and technological achievements [25], and ineffective transformation of results.

Thirdly, there persist weak points in urban governance perceived as obstructions to Guangzhou's primary advancement and developmental objectives. Instances include challenges in urban village transformation, traffic congestion management, and electric bicycle regulation, all requiring more refined and scientifically effective approaches to ensure urban operational safety, production safety, and ecological stability [25]. Consequently, recent alleviation efforts have not sufficiently reduced pressure, highlighting the need to maintain awareness in key areas of risk prevention and mitigation [25]. Fourthly, achieving the ideal of a contented city demands further enhancement in a longer process. While notable progress has been made in childcare and education, there remains ample scope for refinement in this particular area. Local authorities prioritise youth employment and well-being, elderly healthcare, and welfare [7]; however, it is imperative to acknowledge that the masses still contend with numerous daily concerns and livelihood challenges.

Fifthly, there is an expectation for the enhancement of cadre construction and coordination within city governance, surpassing that of other cities in the region, aiming to establish Guangzhou as a national exemplar. However, some officials, in their eagerness to pioneer, may inadvertently compromise their effectiveness, resulting in irregular instances of corruption in crucial sectors. Thus, sustained undertakings are imperative to instigate a lasting transformation in administrative behaviour [25]. Considering these city-level challenges, as well as those stemming from governance and institutional dimensions, city officials have pledged to not only continue advancing solutions but also to intensify efforts, implement appropriate measures, and attain superior outcomes in the progress and development of Guangzhou.

4 Key Lessons Learnt from Guangzhou for Other Cities or Contexts

In this concluding section, we examine several key insights drawn from Guangzhou's development and progress thus far, offering valuable lessons applicable to other contexts and cities. One notable lesson is the significance of historical resilience, as evidenced by Guangzhou's embrace and preservation of its historical roots [58]. This approach not only contributes to the city's identity but also fosters resilience amidst urban development pressures [25]. Guangzhou's economic success is reinforced by its diversified economy [56], cultivated over progressive decades through the encouragement of various industries, enhancing economic stability and growth.

Additionally, Guangzhou's strategic trade location, particularly within developing nations or areas experiencing urban expansion, highlights the importance of strategic positioning in facilitating economic integration and establishing vital trade hubs. Its cultural vibrancy serves as a valuable asset, offering opportunities for prioritising cultural preservation and vibrant public spaces, ultimately enhancing urban life quality and city appeal, a factor less visible in many other Chinese cities.

Furthermore, Guangzhou's tangible global connectivity points out the benefits of international integration, providing insights for cities to leverage geographical positioning for enhanced economic and cultural exchange [60, [35]. The city's innovative urban planning sets a precedent for investing in sustainable urban development and green initiatives [15], contributing to a balanced and liveable urban environment.

Economic collaboration, not only at the city level but also regionally and nationally, is another critical factor [41]. Guangzhou's role in the Greater Bay Area emphasises the importance of economic cooperation [37] between neighbouring regions for mutual growth [53]. Lastly, Guangzhou's success in balancing modernity with tradition demonstrates the feasibility of integrating modern development with historical heritage, offering valuable insights for cities seeking to maintain cultural authenticity amidst economic, sustainable, and/or people-centered urban development plans.

References

1. Allam Z, Cheshmehzangi A, Jones DS (2023a) Urban consumerism and colonial structures of mercantilism. In: Climate and social justice: the political economy of urban resilience and mercantilism, Springer Nature, Singapore, pp 43–60
2. Allam Z, Cheshmehzangi A, Jones, DS (2023b) The economic landscape of cities and climate change. In: Climate and social justice: the political economy of urban resilience and mercantilism, Springer Nature, Singapore, pp 25–41

3. Arase D, de Medeiros Carvalho PMR (eds) (2022) The belt and road initiative in Asia, Africa, and Europe. Taylor & Francis Group, London
4. Bao X (2018) Urban rail transit present situation and future development trends in China: overall analysis based on national policies and strategic plans in 2016–2020. Urban Rail Transit 4(1):1–12
5. Bie J, Jong MD, Derudder B (2015) Greater pearl river delta: historical evolution towards a global city-region. J Urban Technol 22(2):103–123
6. Bui T, Yang D, Jones W, Li J (eds) (2002) China's economic powerhouse: economic reform in Guangdong province, Springer.
7. Chen S, Zhang T (2018) Research on the development and design of rural low-carbon tourism products based on tourist experience taking Xiaozhou village in Guangzhou as an example. In: 3rd international conference on judicial, administrative and humanitarian problems of state structures and economic subjects (JAHP 2018), Atlantis Press, Dordrecht, pp 563–568
8. Chen W, Cheshmehzangi A, Mangi E, Heath T, Ye C, Wang L (2022a) An analysis of residents' social profiles influencing their participation in community micro-regeneration projects in China: a case study of Yongtai community, Guangzhou. Land 11(6):790
9. Chen W, Cheshmehzangi A, Mangi E, Heath T (2022b) Implementations of China's new-type urbanisation: a comparative analysis between targets and practices of key elements' policies. Sustainability 14(10):6341
10. Chen X (2015) A sustainability analysis on the Wuhan-Guangzhou high-speed railway in China. Int J Sustain Transp 9(5):348–363
11. Chen X, Zhu H, Yuan Z (2020) Contested memory amidst rapid urban transition: the cultural politics of urban regeneration in Guangzhou, China. Cities 102:102755
12. Cheshmehzangi A (2016) China's new-type urbanisation plan (NUP) and the foreseeing challenges for decarbonization of cities: a review. Energy Proc 104:146–152
13. Cheshmehzangi A (2020) Towards urban energy management: a brief review of China's recent transition in urban energy demand. Low Carbon Econ 11(01):1
14. Cheshmehzangi A (2021) Low carbon transition at the township level: feasibility study of environmental pollutants and sustainable energy planning. Int J Sustain Energ 40(7):670–696
15. Cheshmehzangi A (2023) Mapping urban regeneration: introduction to the new, the old, and the new-old. In: Mapping urban regeneration: city life experiences in Yunnan, China, Springer Nature, Singapore, pp 1–21
16. Cheshmehzangi A, Chen H (2021a) China's sustainability transitions: low carbon and climate-resilient plan for carbon neutral 2060. Springer Nature, Singapore
17. Cheshmehzangi A, Chen H (2021b) Overview of China's low carbon progress: policies and initiatives. China's sustainability transitions: low carbon and climate-resilient plan for carbon neutral 2060. Springer Nature, Singapore, pp 103–117
18. Cheshmehzangi A, Dawodu A, Sharifi A (2021a) Sustainable urbanism in China. Routledge, New York
19. Cheshmehzangi A, Tang T (2022a) Central plains megalopolis: a hub for green transformation and development beyond just the Beijing-Guangzhou channel. In China's city cluster development in the race to carbon neutrality, Springer Nature, Singapore, pp 147–159
20. Cheshmehzangi A, Tang T (2022b) China's city cluster development in the race to carbon neutrality. Springer Nature, Singapore
21. Cheshmehzangi A, Tang T (2022) Pearl river delta city cluster: from dual-core structure economic development strategies to regional economic plans. In China's city cluster development in the race to carbon neutrality, Springer Nature, Singapore, pp 63–75
22. Cheshmehzangi A, Xie L, Tan-Mullins M (2018) Green development alignments between China and other BRI countries. In: 7th EU-China workshop. April 2018, held in Astana, Kazakhstan
23. Cheshmehzangi A, Xie L, Tan-Mullins M (2021b) Pioneering a green Belt and Road Initiative (BRI) alignment between China and other members: mapping BRI's sustainability plan. Blue-Green Syst 3(1):49–61

24. Cheshmehzangi A, You N, Siri J, Birch E (2024) Identifying four urban innovation-sustainability categories: a summary. In: Harnessing urban innovation to unlock the sustainable development goals. Springer Nature, Singapore, pp 17–29
25. General Office of Guangzhou Municipal People's Government (2024). 2024 Guangzhou municipal government work report <<2024年广州市政府工作报告>> https://www.gz.gov.cn/zwgk/zjgb/zfgzbg/content/post_9462719.html
26. Guangzhou Municipal Government (2021) Notice of the Guangzhou municipal people's government on the issuance of the 14th five-year plan for the national economic and social development of Guangzhou and the outline of long-term goals for 2035. <<广州市人民政府关于印发广州市国民经济和社会发展第十四个五年规划和2035年远景目标纲要的通知. 广州市人民政府.>> https://www.gz.gov.cn/attachment/6/6819/6819148/7339312.pdf
27. He L, Liang Q, Fang S (2016) Challenges and innovative solutions in urban rail transit network operations and management: China's Guangzhou metro experience. Urban rail transit, Vol 2. Springer, Singapore, pp 33–45
28. Ho WC (2021) Globalization, nationalism, and music education in the twenty-first century in greater China. Amsterdam University Press, Amsterdam, p 341 https://www.torrossa.com/gs/resourceProxy?an=4922091&publisher=FZ0675
29. Hou B, Feng P, Hou Z (2013) Planning overall landscapes, building a garden city: on the revision of guangzhou urban green space system planning. J Landsc Res 5(11/12):22
30. Huang L (2006) The future of history: cultural heritage tourism in Guangdong. Doctoral dissertation, University of Technology, Auckland
31. Jiang S (2017) China's Hong Kong: a political and cultural perspective. Springer, Singapore
32. Jiao J, Zhang F, Liu J (2020) A spatiotemporal analysis of the robustness of high-speed rail network in China. Transp Res Part D: Transp Environ 89:102584
33. Kerr G (2013) A short history of China. Oldcastle Books, Harpenden
34. Kielman A (2022) Sonic mobilities: producing worlds in Southern China. University of Chicago Press, Chicago
35. King B, Richards G, Chu AMC (2023) Developing a tourism region through tourism and culture: bordering, branding, placemaking and governance processes. Tourism Recreation Research, 1–15.
36. Lampton DM, Ho S, Kuik CC (2020) Rivers of iron: railroads and Chinese power in Southeast Asia. University of California Press, Oakland
37. Li R, Yu B, Wang Q, Wu G, Ma Z (2024) Changes in economic network patterns and influencing factors in the urban agglomeration of Guangdong-Hong Kong–Macao Greater Bay Area: a comprehensive study. Buildings 14(4):1093
38. Li S (2021) Legal instruments for the integration and cooperation in the Guangdong-Hong Kong-Macao Greater Bay Area (GBA): Better implementation of the SDGs. Sustainability 13(22):12485
39. Lin X, Huang B, Wu J, Ge M, Lin J (2019) Industrial upgrading with innovation production service system: evidence from Guangzhou, China. Procedia CIRP 83:351–357
40. Li S, Haralambides H, Zeng Q (2022) Economic forces shaping the evolution of integrated port systems—the case of the container port system of China's pearl river delta. Res Transp Econ 94:101183
41. Ma S, Ding J, Huang Z, Guo R (2023) Evaluation of the industrial cooperation of the Guangdong Hong Kong Macao Greater Bay Area: based on the origin-destination pairs of industrial parks and coupling model efficiency. PLoS ONE 18(8):e0290572
42. Ma W (2021) From city branding to urban transformation: how do Chinese cities implement city branding strategies. Delft Univ Technol, TPM, Delft
43. Men H (2020) Process of China's open-up. In: China's open-up strategy (1978–2018). Springer, Singapore. https://doi.org/10.1007/978-981-15-4047-9_4
44. Newsgd (2023) Guangdong top 500 manufacturing enterprises list unveiled. https://www.cnbayarea.org.cn/english/News/content/post_1142212.html
45. Ng MK, Xu J (2014) Second metamorphosis? Urban restructuring and planning responses in Guangzhou and Shenzhen in the twenty-first century. In: Maturing megacities: the pearl river delta in progressive transformation, pp 29–60.

46. Shen J, Kee G, Shen J, Kee G (2017) Fuzhou: re-energizing regional economic development and the changing urban functions. In: Development and planning in seven major coastal cities in southern and eastern China, pp 93–118.
47. Shi M, Ma G, Shi Y (2011) How much real cost has China paid for its economic growth? Sustain Sci 6:135–149
48. Song D, Zhang Q (2022) The evolution and driving force on integrating environmental protection and high-quality economic development. Front Econ China 17(3)
49. Southern Metropolis Daily (2022) The high-quality development of Guangzhou metro in the past five years: from a single line network to "one ticket, one city" <<广州地铁高质量发展这五年:从单一线网到"一张票一串城">>, https://m.mp.oeeee.com/a/BAAFRD000020220704699987.html
50. Szto PP (2007) Cultural context and social technology transfer: the case of Canton China. Int J Soc Welf 16(1):55–64
51. Tan-Mullins M, Cheshmehzangi A, Chien S, Xie L (2017) Smart-eco cities in China: trends and city profiles 2016. University of Exeter (SMART-ECO Project), Exeter
52. Tang B, Tan Z (2022) The research on spatial-temporal evolution and influence factors of urban resilience: A case study in the Guangzhou metropolitan area. Front Environ Sci 10:1052930
53. Tang C, Zhang X, Cheung DMW (2024) Nuancing regional geography: an analysis of the evolution and mechanism of intercity cooperation network in the Guangdong-Hong Kong-Macao Greater Bay Area. Tijdschrift voor economische en sociale geografie. https://onlinelibrary.wiley.com/doi/full/10.1111/tesg.12615
54. Teo HC, Lechner AM, Walton GW, Chan FKS, Cheshmehzangi A, Tan-Mullins M, Chan HK, Sternberg T, Campos-Arceiz A (2019) Environmental impacts of infrastructure development under the belt and road initiative. Environments 6(6):72
55. Xu J (2020) Planning intercity railway systems in China's mega-city regions: the case of the pearl river delta. In: Handbook on transport and urban transformation in China. Edward Elgar Publishing, Cheltenham, pp 355–374
56. Xu J, Shao Y (2020) The role of the state in China's post-disaster reconstruction planning: Implications for resilience. Urban Stud 57(3):525–545
57. Yang J, Lin X, Xie Y (2015) Intercity transportation planning in China: case of the Guangzhou-Foshan metropolitan area. Transp Res Rec 2512(1):73–80
58. Yao Z, Li B, Li G, Zeng C (2021) Resilient governance under asymmetric power structure: the case of Enning road regeneration project in Guangzhou, China. Cities 111:102971
59. Zhang H, Tang Y, Yi T (2023) Southeast Asia region. In comparative studies on regional and national economic development. Springer Nature, Singapore, pp 185–277
60. Zhang, J (2018) Intercity relationships within urban agglomeration and their impacts on urban economic development in the case of Guangdong Hong Kong-Macau Greater Bay Area. Intercity relationships within urban agglomeration and their impacts on urban economic development in the case of GuangdongHong Kong-Macau Greater Bay Area, China. MSc Thesis of urban management and development, Erasmus University of Rotterdam, Rotterdam
61. Zhang X (2015) Globalization and the megaregion: investigating the evolution of the Pearl River Delta in a historical perspective. Megaregions. Edward Elgar Publishing, Cheltenham, pp 175–199

Chapter 6
Hangzhou: A Regional Urban Hub for Dynamic Fusion of Tradition and Innovation

1 General Information and Introduction to Hangzhou

Hangzhou, a city with a profound historical legacy, stands as a demonstration to enduring influence and cultural richness in both traditional and contemporary China [68]. Located in the eastern region of China, Hangzhou is not only the capital of Zhejiang Province but also a vital hub of economic and cultural activity [18]. Its historical prominence dates to its role as the capital of the Southern Song Dynasty, a period that greatly shaped its cultural and architectural heritage. Today, Hangzhou seamlessly blends its storied past with contemporary dynamism, which represents its mixed atmosphere of historical and modern areas [59, 60]. Similar to cities of similar scale in China, Hangzhou has gone through vast modernisation processes enabling it to become a new hub for the region and beyond [37]. Hangzhou is characterised by its unique blend of historical significance, natural beauty, and modern dynamism [5, 19]. Its central characteristics are developed based on its economic vitality, cultural richness, modern infrastructure, educational excellence, and tourism [33]. Unlike other cities of Zhejiang Province, Hangzhou plays a more pivotal role in global partnership and international cooperation in multiple industries and sectors.

The city is globally celebrated for the breathtaking beauty of West Lake, a UNESCO World Heritage site [17, 47] that epitomises the harmonious coexistence of natural grandeur and human artistry [6]. Hangzhou's ability to preserve its historical traditions while embracing modern advancements makes it a vital city in China's economic and cultural spheres [15]. Additionally, Hangzhou has emerged as a technology and innovation powerhouse, housing the headquarters of Alibaba Group, one of the world's largest e-commerce companies. This combination of historical richness and modern innovation underscores Hangzhou's unique position as a bridge between China's illustrious past and its promising future. The city has gone through large scale development and redevelopment phases putting it on the map for its novelty in leading innovation in China. With its close proximity to Shanghai, Hangzhou plays a major role in leading regional industries, trade, and businesses. Hangzhou holds a much

A. Cheshmehzangi and T. Tang, *30 Years of Urban Change in China's 10 Core Cities*, Urban Sustainability, https://doi.org/10.1007/978-981-97-8846-0_6

stronger regional value and impact that other cities in the same region, enabling it to be well connected and collaborative with other cities, townships, and rural areas in its vicinity (see Figs. 1–5). It truly remains as an anchor point for Zhejiang Province, through which it contributes to regional development agenda through both (regional) city cluster formation [13] and provincial development.

Recent transformative projects have helped reshape Hangzhou as a major urban and regional hub. For example, in July 2001, the groundbreaking of the Hangzhou Grand Theater marked the official start of the development and construction of Qianjiang New City [20]. This event initiated the city's strategic expansion eastward, tourism development westward, and growth along and across the Qiantang River [27]. This transition signifies a shift from the "West Lake Era" to the "Qiantang

Fig. 1 Hangzhou and its urban development areas in 1985 (map is extracted from open and accessible maps on Google Earth, generated in April 2024)

Fig. 2 Hangzhou and its urban development areas in 1995 (map is extracted from open and accessible maps on Google Earth, generated in April 2024)

Fig. 3 Hangzhou and its urban development areas in 2004 (map is extracted from open and accessible maps on Google Earth, generated in April 2024)

Fig. 4 Hangzhou and its urban development areas in 2014 (map is extracted from open and accessible maps on Google Earth, generated in April 2024)

River Era" and from "cross-river development" to "riverfront development". Over the past 20 years, Hangzhou's urban landscape has been transforming, with the emergence of a vibrant, world-class waterfront area [55–57]. Qianjiang New City has evolved significantly, progressing from its initial phase to a more advanced 2.0 version. Since the commencement of the Hangzhou Grand Theater's construction, the first phase of Qianjiang New City officially opened in October 2008 [27]. This phase unveiled several key features, including the city balcony, the riverside landscape belt, the Xintang River, the new Hangzhou Library, the New Town Tunnel, the Hangzhou Grand Theater, the Civic Center, the International Conference Center, and the Hangzhou Youth Development Center [27]. In December of the same year, the

Fig. 5 Hangzhou and its urban development areas in 2024 (map is extracted from open and accessible maps on Google Earth, generated in April 2024)

Hangzhou East Railway Station hub project was launched, marking the beginning of the Hangzhou East New City's development.

In July 2014, the registration and establishment of the Qianjiang New City Investment Group marked a significant milestone. This event signalled a shift in the development trajectory of Qianjiang New City, adopting a new approach where construction and management functions were separated, as well as management and office operations [27]. The world's attention turned to Qianjiang New City in September 2016, as it hosted the G20 Hangzhou Summit [4], symbolically marking the conclusion of its initial phase, known as "Qianjiang New City 1.0" [27]. In June 2018, the first construction pile was laid for the Asian Games Village, signalling the commencement of development on the south bank of the Qiantang River in preparation for the 19th Asian Games. By December 2020, groundbreaking ceremonies were held for the Hangzhou Grand Convention and Exhibition Center, located at the confluence of the Qiantang River and Hangzhou Bay. This marked the beginning of a new chapter for the Hangzhou Convention and Exhibition New City, as it officially created a new platform for the city [27]. Examples likes this, although not unique only to Hangzhou in the context of China, showcase a chance for reimaging and redeveloping the city's precincts, districts, and particular urban zones (see Figs. 6–9).

1.1 Hangzhou's Physical and Main Development Over the Past 30 years

Hangzhou actively expands the development potential of megacities by embracing comprehensive growth strategies. The city vigorously implements a robust strategy

Fig. 6 Hangzhou's redevelopment and expansion of its central and inner city precincts and zones started in early 1980s, image captured from 1985 (map is extracted from open and accessible maps on Google Earth, generated in April 2024)

Fig. 7 Rapid inner city and central city expansion, image captured from 2004 (map is extracted from open and accessible maps on Google Earth, generated in April 2024)

of urban internationalisation, seeking to enhance its global presence and attract international talent and investment [51, 59, 60]. This strategy includes fostering international business partnerships, cultural exchanges, and tourism initiatives, positioning Hangzhou as a cosmopolitan hub. In parallel, Hangzhou deeply promotes smart city governance, integrating advanced technologies such as big data, artificial intelligence (AI) [53], and the Internet of Things (IoT) to enhance urban management and improve the quality of life for its residents [40]. This commitment to smart governance is reflected in efficient public services [38], intelligent transportation systems, and enhanced public safety measures.

Fig. 8 The introduction of new building masses and large-scale masterplanning projects of mixed-use development, image captured from 2013 (map is extracted from open and accessible maps on Google Earth, generated in April 2024)

Fig. 9 The more formalised expansion of urban areas of mixed-use development, image captured from 2024 (map is extracted from open and accessible maps on Google Earth, generated in April 2024)

The city continuously refines and optimises its functional layout, ensuring that urban development is sustainable and well-planned. This includes developing residential areas, commercial zones, and green spaces in a balanced manner to support economic growth while maintaining environmental sustainability [10]. Through these concerted efforts, Hangzhou has achieved significant results in urban construction [34]. The city's infrastructure has been modernised, creating a more liveable and connected urban environment [28]. Hangzhou's strategic initiatives have not only augmented its economic development but also enhanced its cultural and social fabric, making it a leading example of progressive urban development in China and beyond.

As part of its overarching agenda, major conferences and events are being pushed forward [29], and the internationalisation of the city is being implemented in depth. Over the past ten years, Hangzhou has successfully completed the service guarantee tasks of the G20 Hangzhou Summit [21], successfully bid for and meticulously prepared for the 19th Asian Games and Asian Para Games and introduced the Regulations on Promoting the Internationalisation of Hangzhou City [41]. Focusing on the goal of building a world-class socialist modernised national metropolis, Hangzhou has concentrated on high-level opening up to the outside world, and has been selected as one of the world's 15 sample cities for best practices in tourism and one of the world's top 100 conference destinations, and ranked 80th in the world in terms of the Globalized City Index, and 5th among the nation's large and medium-sized cities [24]. Much of the pathways led to have such status are based on the city's growth and development of these recent three decades.

As the city continues to develop, its spatial layout has been optimised, and the urban areas have expanded to 8289 square kilometers [16]. Over the past ten years, Hangzhou's urban spatial pattern has continuously expanded, and the outline of the city with mountains and water, lake and city, and river has been accelerated. Administrative division orderly adjustments in 2015 and 2017, such as Fuyang and Lin'an, officially withdrawn from the city and set up districts in 2020 as the main city optimisation adjustment [24]. In 2021, the city's urban area expanded to 8289 square kilometers resulting in a more stabilised large-scale urban territories and new development zones of the city [41]. Urbanisation was advanced in an orderly manner, with the urban population increasing from 7,037,000 at the end of 2012 to 10,203,000 at the end of 2021 [24], exceeding 10 million for the first time, and the urbanisation rate increasing from 74.3% to 83.6%, with an average annual increase of 1.03 percentage points.

Hangzhou's multiple transportation hubs have become more convenient, and the rail transportation is connected into a large regional and national network [1]. Over the past ten years, the comprehensive transportation system centered on aviation, high-speed rail, highway, expressway network and subway has been further improved. In 2021, Xiaoshan International Airport opened 268 routes, including 30 international routes, a net increase of 124 and 10 routes respectively compared with 2012 [24]. Air passenger traffic increased from 9.89 million in 2012 to 20.44 million in 2019, but then decreased to 14.32 million in 2021 due to the impact of the pandemic. Hangzhou East Railway Station was reconstructed and put into operation in 2013, and the passenger volume in 2021 became the second highest in China. The total mileage of highways within the city reached 16,989 km, and 801 km of expressways, an increase of 1242 km and 251 km compared with that of 2012 [24]. Rapid development of rail transit, since the first subway line operation in 2012, by the end of June 2022, rail transit operation mileage reached 450 km, ranking 5th in the country's cities [24], in 2021 the total number of passengers 898 million.

In sum, Over the past 30 years, Hangzhou has experienced remarkable physical and developmental transformation, evolving into a dynamic metropolis that balances historical charm with modern innovation. The city's skyline has been significantly altered by the construction of high-rise buildings, including commercial skyscrapers and luxury residential complexes, reflecting its rapid economic growth. Hence, it is now known as a major regional power [43]. The development of infrastructure has been critical to its rapid—yet steady—transformation, with extensive investments in transportation networks such as the Hangzhou Metro system, which now spans multiple lines and connects key areas of the city. Environmental sustainability has also been a priority, with significant efforts to preserve and enhance natural spaces. Aligned with these, the city has prioritised clean energy and green building practices in its urban planning [32]. In addition, Hangzhou's economic development has been robust, driven by the growth of the technology sector, with companies like Alibaba Group headquartered in Hangzhou. This has attracted a talented workforce and fostered a vibrant entrepreneurial ecosystem.

2 Hangzhou's Development Strategies, Including Main Policies and Directions

In more recent years, there is enough evidence to argue that Hangzhou has started to progressively highlight innovation leadership and making every effort to shape new advantages in development through scientific and technological innovation. The "315" science and technology innovation system construction project has been implemented in depth, and new industries, new modes and new kinetic energy have been initiated by subversive and cutting-edge technologies [26]. Through various effective ways of accelerating the construction of a global innovation center, Hangzhou continues to promote the high-quality integration and development of the West Science and Innovation Corridor, and actively build a matrix of strategic scientific and technological forces. In recent years, the city has accelerated the construction of national laboratories, and strive to put into operation a large scientific device, start a project, and plan a project, and make breakthroughs in the Deep Time Digital Earth International Big Science Program [26]. In doing so, it aims to consolidate the foundation for the creation of a comprehensive national science center.

Hangzhou is prioritising advancements in cutting-edge fields such as chip technology, cloud computing, artificial intelligence, and key biotechnologies. The city is actively involved in implementing over 100 major research and development projects at the national, provincial, and city levels, with a goal of achieving significant breakthroughs and replacing over 30 key items with innovative solutions [26]. To support these ambitious projects, Hangzhou strictly adheres to the "two 15%" policy, which ensures that a substantial portion of resources is allocated efficiently. The city is also improving the evaluation mechanisms for the distribution of scientific research funds to maximise their impact [26]. This strategic approach involves investing a considerable 80 billion yuan in research and development across the society, accentuating Hangzhou's commitment to fostering a robust innovation ecosystem. Moreover, the city's focus on intellectual property is evident in its impressive achievement of owning 48 high-value invention patents per 10,000 residents [26]. This emphasis on patent development highlights Hangzhou's dedication to leading in technological innovation and securing its position as a major player in the global tech landscape [44]. Through these comprehensive efforts, Hangzhou aims to drive significant progress in science and technology, contributing to both local and national development goals.

Furthermore, Hangzhou highlights the integration of numbers and realities, and making every effort to build a modernised industrial system [66]. In-depth implementation of critical projects, such as digital economy innovation and quality "one development project" [7], suggest new pathways to promote the "415X" advanced manufacturing cluster cultivation and high-quality development of the service industry [26]. Such approaches help to accelerate the construction of the digital economy as the core [14], the real economy as the support, scientific and technological innovation as the lead of modernised industrial system [26]. In Hangzhou, the modernised industrial system is led by science and technology innovation [68], opening up new space and expansion opportunities for further economic and physical development in the city and its region.

As part of its vision for future growth, Hangzhou is concentrating on boosting demand across various sectors [54]. The local authorities are committed to ensuring the economy remains stable and operates within a sustainable range [12]. To achieve this, they emphasise prioritising key projects as central to their strategy. This includes accelerating consumer spending recovery, strengthening foreign trade, and attracting foreign investment [31]. The plan involves a multifaceted approach: promoting large-scale projects that drive economic activity, enhancing policies to stimulate domestic consumption [62], and creating a favourable environment for international business and investment [26]. By consolidating foreign trade and investment, Hangzhou aims to maintain robust economic connections with global markets, ensuring a steady inflow of capital and resources. The overarching goal is to achieve high-quality economic improvements alongside reasonable quantitative growth [26]. This means not only increasing the volume of economic activities but also enhancing the efficiency and value of these activities. By focusing on strategic projects, boosting consumer confidence and spending, and reinforcing its position in international trade, Hangzhou aims to ensure sustainable and balanced economic development that benefits both its residents and its business community [50]. This comprehensive approach is designed to foster long-term prosperity and resilience in Hangzhou's economy.

In addition to above, Hangzhou continues to highlight the reforms to break through the difficulties and make every effort to build the city with the best business environment [55, 57] and the strongest private economy [39]. In doing so, it has worked rigorously on highlighting functional quality [26]. Thus, the authorities have pledged to make every effort to solve the problems of mega-city governance and adhere to systematic thinking through strengthening digital intelligence. The overall plan is to accelerate urban organic renewal, and deeply explore new ways of governance in line with the characteristics and laws of mega-cities.

One of the key areas of focus is also to optimise the spatial pattern of cities. By concentrating on the nature of the city and the core functional position, the city aims to enhance urban recognition and accelerate the development of strategic key blocks [26]. In its major development plans, the city is forecasted to focus on the west of the city science and innovation corridor, the east of the city wisdom corridor, Liangzhu cultural corridor, as well as areas along the Qiantang River Financial Harbor 4 corridors [26]. In delivering such major projects and initiatives, the city aims to promote the future of science and technology city, the cloud city, the demonstration zone of the pro-airport economy, the new district of Qiantang, the second phase of the new city of the Qianjiang river, the Qianjiang Century City, the big city of the north, the south of the river, and the central zone of the city. The city also includes a vast range of smart city initiatives in other areas of fast construction, fast completion, and developing them into a piece of momentum [26]. The smart city initiatives in Hangzhou have developed as a major capacity to create a new landmark in the city.

3 Hangzhou Future Sci-Tech City

A recent exemplary urban transition project in Hangzhou is the development of the Hangzhou Future Sci-Tech City. This project exemplifies Hangzhou's commitment to integrating advanced technology with sustainable urban planning [2]. The Hangzhou Future Sci-Tech City, also known as the Cloud Town, is an ambitious urban development project that aims to create a modern, high-tech urban district. It focuses on fostering innovation and accommodating the rapid growth of the technology sector in Hangzhou. The project fits well in the Hangzhou West Hi-Tech Corridor, where three major 'Science and Technology City' hubs are located. These include Qingshanhu Science and Technology City in the west, Future Science and Technology City in the middle, and Zijingang Science and Technology City in the east. The link between them is through the Science and Technology Avenue and Wenyi West Road (see Figs. 10–12). In between Future Sci-Tech City and Zijingang Sci-tech City, there sits the Cloud City zone, which is the newest addition to this extended Sci-Tech corridor.

The Hangzhou Future Sci-Tech City project began taking shape in the early 2010s, with the aim of transforming Hangzhou into a leading global hub for innovation and technology [67]. While the precise start date may vary depending on the specific phases of development, key milestones indicate the gradual progression of this ambitious initiative. One significant milestone in the development of the Hangzhou Future Sci-Tech City was the announcement of plans for the project in the mid to late 2000s [3]. During this time, discussions began regarding the creation of a dedicated area in Hangzhou that would focus on fostering technological innovation and attracting high-tech industries. Construction and development activities gained momentum in the early 2010s, with initial groundwork and infrastructure projects laying the foundation for future growth [42]. This period saw the establishment of key components such as research institutions, innovation centers, and strategic partnerships with leading technology companies.

By the mid-2010s, the Hangzhou Future Sci-Tech City started to take shape more visibly, with the completion of landmark buildings and the establishment of anchor

Fig. 10 Hangzhou's West Hight-Tech Corridor from 1987 to 2024, with legend on 2004 map indicating location of new development zones for four main city areas and the corridor between them from Keji Avenue in the west to Wenyixi Road in the east. Map is from 1987. (map is extracted from open and accessible maps on Google Earth, generated in April 2024)

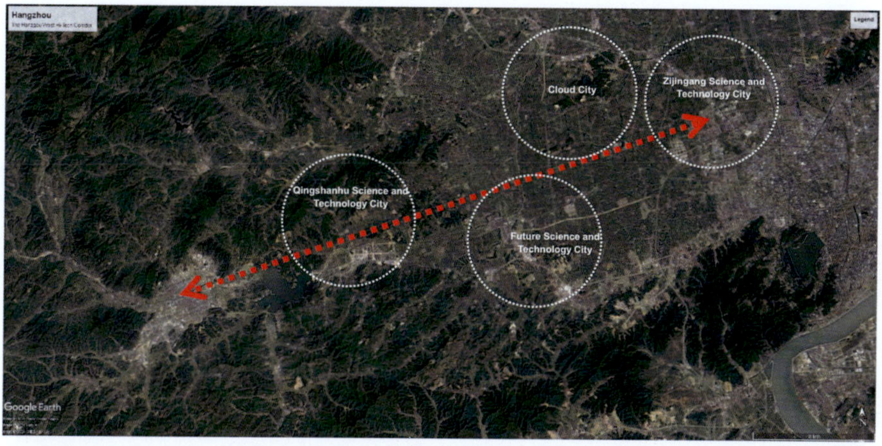

Fig. 11 Hangzhou's West Hight-Tech Corridor from 1987 to 2024, with legend on 2004 map indicating location of new development zones for four main city areas and the corridor between them from Keji Avenue in the west to Wenyixi Road in the east. Map is from 2004. (map is extracted from open and accessible maps on Google Earth, generated in April 2024)

tenants within the area. This period also saw the implementation of smart city initiatives and the integration of advanced technologies into urban infrastructure. As the project progressed, it garnered increased attention and investment, both domestically and internationally [61]. The Hangzhou Future Sci-Tech City became a focal point for discussions on urban development, innovation ecosystems, and the future of technology-driven economies. Overall, while there may not be a single definitive start date for the Hangzhou Future Sci-Tech City project [14], its development has been characterised by a series of strategic initiatives and progressive milestones that have positioned Hangzhou as a leading center for innovation and technology in China and beyond.

A central feature for this area is its technological hub characteristics. The city is designed to host a plethora of tech companies, research institutions, and innovation centers [23]. It provides an ecosystem conducive to tech startups and established tech giants alike, with Alibaba's headquarters being a foundation of this development. It also includes an extensive smart infrastructure [45]. The project integrates smart city technologies, including IoT, big data, and artificial intelligence to enhance urban management [35]. This includes intelligent traffic management systems, smart energy grids, and advanced waste management solutions. All of these also link to three key areas of sustainable development (including sustainable transportation and buildings), education and talent development, and the development of cultural and recreational facilities.

In three specific areas, the Hangzhou Future Sci-Tech City has had impact on its residents and wider community [36]. These include impacts on economic growth, urban liveability, and global connectivity [45]. By attracting high-tech businesses and fostering innovation, the Hangzhou Future Sci-Tech City significantly contributes to the economic growth of the region. It serves as a catalyst for the development of the digital economy and high-tech industries. More importantly, the project enhances urban liveability through its focus on sustainable practices, smart infrastructure, and comprehensive amenities [48]. This makes Hangzhou an attractive place for both living and working. Lastly, the project strengthens Hangzhou's position on the global stage as a leader in technological innovation and sustainable urban development.

Fig. 12 Hangzhou's West Hight-Tech Corridor from 1987 to 2024, with legend on 2004 map indicating location of new development zones for four main city areas and the corridor between them from Keji Avenue in the west to Wenyixi Road in the east. Map is from 2024. (map is extracted from open and accessible maps on Google Earth, generated in April 2024)

Therefore, the Hangzhou Future Sci-Tech City stands as a prime example of how urban development can be driven by technological innovation [49], sustainability, and strategic planning, ensuring the city's growth and adaptability in the modern era.

4 The Local Government Projection for Hangzhou's Future Development

As the local authorities highlight very openly, the comprehensive capacity, core competitiveness, and international reputation of the city are planned to be greatly improved. These will be aligned well with the overall approach to foster the enhancement of the level of modernisation of urban governance [30]. In fast-forwarding the modernisation of urban governance, the next steps are to basically build a modern socialist international metropolis. This is based on the fact that the city continues to capitalise on its solid foundation for the construction of a globally influential world city with unique flavour and unique splendour by the middle of this century [25]. It will then progress to fully demonstrate the superiority of the socialist system with Chinese characteristics.

The city officials also develop this overarching narrative that Hangzhou will take the lead in promoting innovation-driven development through digital transformation, entering the forefront of innovative cities, and building a city that meets the requirements of high-quality development [22]. This will be based on a so-called modernised economic system that meets the requirements of high-quality development, and exploring the formation of a new development pattern that will effectively contribute to the construction of a new development pattern. In doing so, effective paths will be conducive to the construction of a new development pattern (i.e., partly through the city's multiple new development zones), meanwhile the city's innovative capacity and comprehensive strength will reach a new level. While this is somewhat expected as a by-default impact of such development, we anticipate seeing increase in the per capita GDP that could reach the level of developed economies.

Taking the lead in promoting the construction of a new type of smart city [58] with a city brain, Hangzhou is comprehensively building an overall smart governance system [52]. This initiative aims to elevate the construction of Rule of Law Hangzhou and Safe Hangzhou to a higher level, while perfecting the pattern of shared social governance [63]. Demonstrating cultural self-confidence, Hangzhou showcases its unique cultural charms, significantly enhancing the soft power and influence of urban culture [64]. This venture not only improves the modernisation of individuals but also elevates the civilization of society. Moreover, Hangzhou leads the way in promoting common prosperity through urban and rural regional integration, effectively addressing the problem of unbalanced and insufficient development [25]. This includes initiatives to enhance the quality of education, healthcare, and social welfare, fostering a more equitable and inclusive society.

The quality of public services, including education, health, and sports, has seen significant improvements, resulting in greater equality and enhanced standards of living for residents. Education, healthcare, sports, and other essential services have been elevated to higher levels of quality and equity, leading to continuous enhancements in people's quality of life [25]. This upward trajectory in living standards is emphasised by the ongoing optimisation of urban and rural habitats, with a concerted focus on transforming green landscapes into prosperous economic assets through extensive garden construction initiatives [10]. Through these efforts, the city has achieved remarkable results in green development and ecological enrichment [11], culminating in the creation of a modern urban environment reminiscent of the idyllic scenery depicted in the ancient Chinese painting "Dwelling in the Fuchun Mountains" [25]. Anchored in the visionary goals set for 2035, Hangzhou remains committed to problem-oriented, integrative, and innovative approaches to economic and social development, with a persistent focus on achieving key objectives over the next five years.

The overarching goals for Hangzhou are to emerge as a leader in comprehensive strength, with a focus on robust economic development and the construction of a modernised economic system. This entails sustaining a healthy and rapid pace of economic growth while harnessing the full potential of growth and significantly enhancing innovation capacity. Specifically, the city aims to surpass key economic benchmarks, with the gross domestic product (GDP) exceeding 2.3 trillion yuan and per capita GDP surpassing 180,000 yuan. Additionally, the city aspires to bolster its research and experimentation (R&E) capacity, aiming for it to exceed 1.5 trillion yuan. To achieve these targets, efforts will be concentrated on increasing the intensity of investment in research and experimental development to reach four per cent, thereby fuelling innovation, and driving sustainable economic progress [25]. Through these rigorous endeavours, Hangzhou is prepared to reinforce its position as a powerhouse of economic vitality and innovation on both regional and global scales.

At the forefront of Hangzhou's development agenda is the prioritisation of digital transformation, with a focus on seamlessly integrating digital industrialisation, industrial digitisation, and urban digitisation [25]. This comprehensive approach aims to drive innovation across various sectors, promoting the development of new infrastructure, new consumption patterns, new manufacturing processes, new e-commerce models, and advancements in healthcare. Concurrently, efforts are underway to cultivate a world-class business environment conducive to international investment and trade. Simultaneously, cultural construction takes center stage, with a strong emphasis on promoting socialist core values and preserving historical heritage. The spirit of the Red Boat and the essence of Zhejiang's cultural identity are actively promoted, while initiatives to elevate Hangzhou as a famous historical and cultural city are ongoing. Through these accomplishments, the city aims to foster a culture of innovation and urbanisation, exemplifying its commitment to progress and heritage preservation. Moreover, Hangzhou's cultural significance is further underscored by the successful hosting of the 2022 Hangzhou Asian Games and Asian Para Games, highlighting the city's ability to showcase its rich cultural heritage on an international stage. Looking ahead, Hangzhou endeavours to strengthen its public cultural service system and cultural industry framework, solidifying its status as an important hub for international exchange of oriental culture and a thriving center for cultural and creative activities. Through these multifaceted efforts, Hangzhou is poised to emerge as a global leader in both technological innovation and cultural excellence.

Hangzhou, known for its rich cultural heritage and technological innovation, is also committed to preserving its ecological environment. At the forefront of this attempt is the transformation of green water and green mountains into assets of immense value, further broadening the scope of ecological conservation efforts [25]. The city has implemented a comprehensive ecological civilization system, fostering a harmonious balance between environmental sustainability and socio-economic development.

Efforts to promote green transformation encompass various aspects of production and lifestyle, with initiatives aimed at reducing carbon footprint and promoting sustainable practices across industries [8, 9]. Notably, the integration of protection and enhancement measures for water bodies such as the West Lake, West Stream, Qiandao Lake, Xin'an River, Fuchun River, Qiantang River, Campsis Creek, and the Grand Canal exemplify Hangzhou's commitment to preserving its natural resources [25]. As a result of these determined efforts, the ecological environment of these important water systems has significantly improved, with noticeable enhancements in their beauty and ecological integrity. Hangzhou has embraced the concept of a "wetland and water city", where the coexistence of natural ecosystems and urban development is celebrated. This distinctive feature highlights Hangzhou's dedication to sustainability and sets it apart as a model city in the new era.

At the core of Hangzhou's development agenda is the prioritisation of enhancing the quality of life for its residents. With per capita disposable income exceeding 85,000 yuan and the urban–rural income gap narrowing to less than 1.8 yuan, people's well-being has seen significant improvements [25], illustrating the city's regional and city-level directions are tangible. Furthermore, Hangzhou is committed to meeting the needs of its residents across all stages of life, with a focus on elevating education and healthcare standards. As a result, per capita life expectancy is expected to rise to 83.88 years, and the urbanisation rate of the resident population will surpass 82 per cent [25]. These efforts are complemented by initiatives to enhance the quality and balance of public services, all aimed at fostering a sense of happiness and well-being among the city's inhabitants, ultimately striving to realise the vision of a "Happy City".

4.1 Hangzhou's Challenges of the Past and Present

The current external economic landscape presents multifaceted challenges, necessitating a comprehensive approach to address various facets of economic stability and growth. There is heightened pressure to stabilize exports amidst a complex and severe global environment. Furthermore, inadequate effective demand coupled with sluggish consumption patterns underscores the need for strategic interventions to stimulate economic activity. Social expectations remain subdued, contributing to tepid private investment sentiments, particularly in key sectors [26]. This subdued economic sentiment is particularly evident in the struggles faced by certain industries and small to medium-sized micro-enterprises in maintaining viable production and operations. Despite advancements in scientific and technological innovations, the anticipated growth of new kinetic energy sources has been sluggish, with insufficient synergistic innovation and transformation across industry, academia, and research institutions. This lack of collaborative momentum poses risks to the development of certain fields, highlighting the importance of fostering stronger interdisciplinary collaborations. Moreover, the nascent industrial tracks are still in the cultivation phase, lacking major projects to serve as catalysts for explosive growth. People's livelihoods continue to exhibit numerous shortcomings and weaknesses, posing risks across various sectors such as the economy, finance, safety, and production [26]. Additionally, a subset of government personnel lacks the requisite professional competence and commitment to effectively address emerging challenges and align with evolving societal needs and expectations. To effectively navigate these challenges, a targeted and proactive approach is imperative, wherein strategic measures are implemented to address each area of concern comprehensively, fostering sustainable economic growth and societal well-being.

5 Key Lessons Learnt from Hangzhou for Other Cities and Regions in China and Globally

Key lessons learned from Hangzhou provide valuable insights for other cities aiming to achieve sustainable development and economic prosperity:

Harmonising Tradition and Modernisation: Hangzhou has demonstrated the importance of successfully preserving cultural heritage while embracing modernization. By striking a balance between tradition and innovation, cities can maintain their unique cultural identity while adapting to the demands of a rapidly evolving world.

Sustainable Tourism Practices: Hangzhou's approach to tourism growth emphasises environmental preservation, particularly for UNESCO sites like the West Lake. Other cities can learn from Hangzhou's example by implementing sustainable tourism practices that prioritise the protection of natural and cultural resources for future generations.

Strategic Infrastructure Investment: Hangzhou's strategic investment in infrastructure has been pivotal for economic growth and urban expansion. By prioritising key projects that enhance connectivity and facilitate economic development, cities can create a foundation for sustainable growth and prosperity.

Efficient Transportation Hubs: Hangzhou's efficient transportation hubs, including well-planned airports and railway stations, have played a crucial role in enhancing connectivity and facilitating economic activities. Other cities can emulate Hangzhou's approach by investing in well-designed transportation infrastructure to improve accessibility and mobility for residents and visitors alike.
Innovation Hub Planning: Hangzhou's focus on technology and innovation has propelled it to become a global hub for innovation. Cities looking to foster innovation ecosystems can learn from Hangzhou's success by creating conducive environments that attract talent, foster collaboration, and support entrepreneurship.
Cultural Preservation in Urban Development: Hangzhou has integrated historical waterways and cultural spaces into its urban development, preserving its cultural heritage while promoting sustainable growth. By incorporating cultural preservation into urban planning strategies, cities can create vibrant and culturally rich environments that enhance quality of life for residents.
Green and Sustainable Urban Environments: Hangzhou's prioritisation of green spaces has contributed to enhanced liveability and environmental sustainability. Cities can follow Hangzhou's lead by prioritising green infrastructure and sustainable urban planning practices to create healthier and more resilient communities.
Public–Private Collaboration: Hangzhou's successful partnerships between the government and private sector have been instrumental in driving efficient development. By fostering collaboration and cooperation between public and private entities, cities can leverage resources and expertise to address complex urban challenges and achieve shared goals of sustainable development and prosperity.

In conclusion, Hangzhou serves as a beacon of inspiration for cities worldwide seeking to achieve sustainable development and economic prosperity [65]. Through its successful harmonisation of tradition and modernisation, strategic infrastructure investments, and emphasis on innovation and sustainable practices, Hangzhou has emerged as a model city for urban development. By prioritising cultural preservation, green spaces, and efficient transportation, Hangzhou has created a vibrant and liveable urban environment that prioritises the well-being of its residents [46]. Furthermore, Hangzhou's effective public–private collaborations emphasise the importance of partnerships in driving efficient and inclusive development. As cities navigate the complexities of urbanisation and globalisation, the lessons learned from Hangzhou provide invaluable guidance for similar regional urban hubs in China and other contexts/countries.

References

1. An Y, Wei YD, Yuan F, Chen W (2022) Impacts of high-speed rails on urban networks and regional development: a study of the Yangtze River Delta, China. Int J Sustain Transp 16(6):483–495

2. Anzoise V, Slanzi D, Poli I (2020) Local stakeholders' narratives about large-scale urban development: the Zhejiang Hangzhou future sci-tech city. Urban Stud 57(3):655–671
3. Argyriou I (2019) The smart city of Hangzhou, China: the case of Dream Town Internet village. Smart city emergence. Elsevier, Amsterdam, pp 195–218
4. Bradford CI (2016) G20 Hangzhou Summit: a possible turning point for global governance. China Quart Int Strateg Stud 2(03):327–346
5. Cao K, Xie W, Zhu J, Wei F (2023) Commercial culture as a key impetus in shaping and transforming urban structure: case study of Hangzhou, China. Sustainability 15(13):10620
6. Cheshmehzangi A (2020) Identity and the city. In: Identity of cities and city of identities. Springer, Singapore, pp 15–83
7. Cheshmehzangi A (2021) From transitions to transformation: a brief review of the potential impacts of COVID-19 on boosting digitization, digitalization, and systems thinking in the built environment. J Build Constr Plan Res 9(1):26–39
8. Cheshmehzangi A, Chen H (2021a) China's sustainability transitions: low carbon and climate-resilient plan for carbon neutral 2060. Springer Nature, Singapore
9. Cheshmehzangi A, Chen H (2021b) Overview of China's low carbon progress: policies and initiatives. China's sustainability transitions: low carbon and climate-resilient plan for carbon neutral 2060. Springer Nature, Singapore, pp 103–117
10. Cheshmehzangi A, Dawodu A, Sharifi A (2021a) Sustainable urbanism in China. Routledge, New York
11. Cheshmehzangi A, Flynn A, Tan-Mullins M, Xie L, Deng W, Mangi E, Chen W (2021b) From eco-urbanism to eco-fusion: an augmented multi-scalar framework in sustainable urbanism. Sustainability 13(4):2373
12. Cheshmehzangi A, Dawodu A, Song W, Shi Y, Wang Y (2020) An introduction to neighborhood sustainability assessment tool (NSAT) study for China from comprehensive analysis of eight Asian tools. Sustainability 12(6):2462
13. Cheshmehzangi A, Tang T (2022a) China's city cluster development in the race to carbon neutrality. Springer Nature, Singapore
14. Cheshmehzangi A, Tang T (2022b) Yangtze River Delta City Cluster: from integrated production and services to digital technology innovation. In China's city cluster development in the race to carbon neutrality, Springer Nature Singapore, Singapore, pp 47–62
15. Chong Z, Qin C, Ye X (2016) Environmental regulation, economic network and sustainable growth of urban agglomerations in China. Sustainability 8(5):467
16. Dai T, Yan H, Bao Z, Wu R, Nan X, Luo Y, Yang M (2022) Have residents changed their perceptions and use of community park green space during the covid-19 pandemic?
17. Feng H (2015) Historic urban landscape research in China: The Slender West Lake in Yangzhou. Hist Environ 27(1):68–77
18. Forster K, Yap X (2005) A comparative analysis of economic reform and development in Hangzhou and Wenzhou cities. Cities in Post-Mao China. Routledge, New York, pp 51–102
19. Fu S (2016) Shan-shui myth and history: the locally planned process of combining the ancient city and West Lake in Hangzhou, 1896–1927. Plan Perspect 31(3):363–390
20. Fu S, Jin P (2022) Urban morphological changes in Hangzhou from the 1950s to the 1990s. In: Cities as assemblages: Proceedings of the XXVI International Seminar on Urban Form 2019, 2–6 July 2019, Nicosia, Cyprus. http://www.tabedizioni.com/camiz/capitoli_vol_3/978889295357425.pdf
21. Gao J, Jia R, Su Q (2022) G20 entrepreneurship services report. Palgrave Macmillan, Singapore
22. Han Y, Cai J, Ma E, Du S, Lin J (2023) Understanding smart city practice in urban China: a governance perspective. Sustainability 15(9):7034
23. Han Y, Du X, Zhang H, Ni J, Fan F (2024) Does smart home adoption reduce household electricity-related CO_2 emissions? Evidence from Hangzhou city, China. Energy 289:129890
24. Hangzhou Municipal Bureau of Statistics (2022) Forge ahead into the new era and build a new paradise: a few statistics on the past ten years in Hangzhou. <<奋进新时代 建设新天堂—数说杭州这十年>>. http://tjj.hangzhou.gov.cn/art/2022/9/20/art_1229279240_4088462.html

25. Hangzhou Municipal People's Government Portal (2021) The 14th Five-Year Plan for national economic and social development of Hangzhou City and the outline of long-term goals for 2035 <<杭州市国民经济和社会发展第十四个五年规划和二〇三五年远景目标纲要>>. https://zjjcmspublic.oss-cn-hangzhou-zwynet-d01-a.internet.cloud.zj.gov.cn/jcms_files/jcms1/web149/site/attach/0/d9c52dba7b7545be823485f5b8f6f100.pdf
26. Hangzhou Municipal People's Government Portal (2024) 2024 Hangzhou Municipal Government Work Report <<2024年杭州市政府工作报告>>. https://www.hangzhou.gov.cn/art/2024/2/23/art_1229063401_4241383.html
27. Hangzhou Net (2021) From tidal flats to the city's CBD, a forum detailed the 20-year "city-building" history of Qianjiang New Town and looked forward to the "leapfrog" path of the 2.0 new era <<从滩涂到城市CBD,一场论坛细说钱江新城20年"建城"史 展望2.0新时代"跃变"路>>. https://ori.hangzhou.com.cn/ornews/content/2021-07/10/content_8005189.htm
28. Hassenpflug D, Kammerbauer M (2010) The urban code of China. Birkhäuser, Basel
29. Hua G (2022) International conference destination competitiveness: measures and promotion strategies for Hangzhou. Forest Chem Rev 1109–1128.
30. Huang JW, Teng W, Galliers R (2016) Transitioning beyond the economic-environmental trade-off in the city of Hangzhou. In: 2016 Annual Meeting of the Academy of Management, Anaheim, California, USA, 5–9 August 2016. https://irep.ntu.ac.uk/id/eprint/29703/
31. Jiang B, Ding L, Fang X (2019) Sustainable development of new urbanization from the perspective of coordination: a new complex system of urbanization-technology innovation and the atmospheric environment. Atmosphere 10(11):652
32. Kong D, Cheshmehzangi A, Zhang Z, Ardakani SP, Gu T (2023) Urban building energy modeling (UBEM): a systematic review of challenges and opportunities. Energ Effi 16(6):69
33. Lam JS (2017) Senses of the city: perceptions of Hangzhou and Southern Song China, 1127–1279. The Chinese University of Hong Kong Press, Hong Kong
34. Li H, Cheshmehzangi A, Zhang Z, Su Z, Pourroostaei Ardakani S, Sedrez M, Dawodu A (2022) The correlation analysis between air quality and construction sites: evaluation in the urban environment during the COVID-19 pandemic. Sustainability 14(12):7075
35. Marvin S, While A, Chen B, Kovacic M (2022) Urban AI in China: social control or hyper-capitalist development in the post-smart city? Front Sustain Cities 4:1030318
36. Miao JT, Phelps NA, Lu T, Wang CC (2019) The trials of China's technoburbia: the case of the future Sci-tech City corridor in Hangzhou. Urban Geogr 40(10):1443–1466
37. Mohiuddin M, Regnière MH, Su A, Su Z (2014) The special economic zone as a locomotive for green development in China. Asian Soc Sci 10(18):109
38. Pourroostaei Ardakani S, Cheshmehzangi A (2023) Moving forward with big data analytics and smartness. In: Big data analytics for smart urban systems. Springer Nature Singapore, Singapore, pp 111–126
39. Shen J, Kee G (2017) Hangzhou: raising the urban competitiveness with new socio-economic strategies and spatial roles. In: Development and planning in seven major coastal cities in southern and eastern China, pp 153–180.
40. Tan-Mullins M, Cheshmehzangi A, Chien S, Xie L (2017) Smart-eco cities in China: trends and city profiles 2016. University of Exeter (SMART-ECO Project), Exeter
41. The Supreme People's Procuratorate of the People's Republic of China (2022) Forge ahead on a new journey and build a new era of meritorious service—follow the footsteps of the General Secretary to see changes | Hangzhou, Zhejiang: Actively build a "global digital rule of law supervision system" to promote the modernization of municipal governance <<奋进新征程 建功新时代——沿着总书记的足迹看变化|浙江杭州:积极构建"全域数字法治监督体系"推动市域治理现代化>> https://www.spp.gov.cn/zdgz/202205/t20220525_557715.shtml
42. Wang HH, Miao ML (2024) The future of China's development and globalization: views from ambassadors to China. Springer Nature, Singapore, p 222
43. Wang J, Chen T, Chen Z, Lin Y (2023) Building tall, falling short: an empirical assessment of Chinese skyscrapers. Falling short: an empirical assessment of Chinese skyscrapers. https://papers.ssrn.com/sol3/papers.cfm?abstract_id=4462688

44. Wang J, Dong W (2022) Multiple effects of urban innovation activities on growth of key industries: case study in Hangzhou, China. J Urban Plan Dev 148(1):05021054
45. Wang W, Liang J, Niu J (2022) Site selection of co-working spaces under the influence of multiple factors: a case study in Hangzhou, China. Sustainability 14(5):2676
46. Wang Y (2020) Data-driven smart mobility as an act to mitigate climate change, a case of Hangzhou. https://www.diva-portal.org/smash/record.jsf?pid=diva2%3A1437243&dswid=9683
47. Wen W, Keane M (2014) Developing creative service industries in Hangzhou, People's Republic of China [Tourism and the creative economy (OECD Studies on Tourism)]
48. Wu Y, Yang Y, Chen Q, Xu W (2020) The correlation between the jobs-housing relationship and the innovative development of Sci-Tech parks in new urban districts: a case study of the Hangzhou West Hi-Tech Corridor in China. ISPRS Int J Geo Inf 9(12):762
49. Wu Y, Li X, Yang Y, Xu W (2022) Correlation between urban commercial nodes and the development of sci-tech enterprises in Hangzhou West High-Tech Corridor, China. Land 11(11):2086
50. Xia Y (2023) Impact of green space on residents' wellbeing: acase study of the Grand Canal (Hangzhou section). Front Public Health 11:1146892
51. Xu H (2022) The road map of the rise of national central cities under the Belt and Road initiative. In: The Belt and Road: industrial and spatial coordinated development, pp 219–242. Springer Nature Singapore, Singapore.
52. Yan X, Li T (2023) Construction and application of urban digital infrastructure—practice of "urban brain" in facing COVID-19 in Hangzhou, China. Eng Constr Archit Manag 30(8):3123–3141
53. Yan Z, Jiang L, Huang X, Zhang L, Zhou X (2023) Intelligent urbanism with artificial intelligence in shaping tomorrow's smart cities: current developments, trends, and future directions. J Cloud Comput 12(1):179
54. Yang Y, Chen H, Zhang Q, Lei J (2018) The commercialization of university and research institutes' science-based innovations: the four successful Chinese cases. J Ind Integr Manage 3(03):1850013
55. Yin J, Fu P, Cheshmehzangi A, Li Z, Dong J (2022a) Investigating the changes in urban green-space patterns with urban land-use changes: a case study in Hangzhou, China. Remote Sens 14(21):5410
56. Yin J, Fu P, Hamm NA, Li Z, You N, He Y, Cheshmehzangi A, Dong J (2021) Decision-level and feature-level integration of remote sensing and geospatial big data for urban land use mapping. Remote Sens 13(8):1579
57. Yin JH, Song HY, Zeng KX (2022b) Does a smart business environment promote corporate investment? a case study of Hangzhou. PLoS ONE 17(7):e0269089
58. Yin M (2021) A systematic review of smart city development policies in Chinese cities. Chulalongkorn University Theses and Dissertations (Chula ETD), 4933. https://digital.car.chula.ac.th/chulaetd/4933
59. Zhang H (2023) The renewal strategies of waterfront industrial heritages in Guangzhou from an urban catalyst perspective—a case study on Wenchong Shipyard, Doctoral dissertation, Politecnico di Torino.
60. Zhang Q (2023) Geographic distribution and location choice of foreign direct investment in China, Doctoral dissertation, Queen's University Belfast.
61. Zhang M, Xu J (2024) Central–local state strategic coupling towards making entrepreneurial cities in China. Polit Geogr 109:103062
62. Zhang Z, Cheshmehzangi A, Ardakani SP (2021) A data-driven clustering analysis for the impact of COVID-19 on the electricity consumption pattern of Zhejiang Province, China. Energies 14(23):8187
63. Zhao H, Chen H, Zhao H, Lu Y, Zhu J (2023) What business environment element configuration can promote urban digitization development? Environ Dev Sustain 1–22
64. Zhou Z, Lee J (2021) Government-enterprise cooperation under the construction of smart cities in Hangzhou, China. In: 2021 International conference on public management and intelligent society (PMIS), pp 234–237. IEEE.

65. Zou X (2023) Urban sustainable development in East Asia: understanding and evaluating urban sustainable trends in China and Japan. Springer Nature, Singapore
66. Zhu K, Xu J, Wang X (2023a) The evolution of urban innovation space and its spatial relationships with talents' living demands: evidence from Hangzhou, China. Int J Urban Sci 27(3):442–460
67. Zhu K, Zhang F, Wu F (2023b) Creating a state strategic innovation space: the development of the Zhangjiang Science City in Shanghai. Int J Urban Sci 27(4):599–621
68. Zhu W, Chen J (2022) The spatial analysis of digital economy and urban development: a case study in Hangzhou, China. Cities 123:103563

Chapter 7
Chengdu: From Regional Ascendance to Integrating a Regional Innovation Ecosystem

1 Introduction to Chengdu and Its Development

Chengdu, the capital of Sichuan Province, is one of China's 15 sub-provincial cities. Strategically located on the western edge of the Sichuan Basin, Chengdu spans an area of 14,335 square kilometers. It serves as the commercial logistics center and a comprehensive transportation hub for western China [61]. Its position at the geographic center of southwest China provides it with significant geographical advantages. Situated at the midpoint of the Eurasian route, Chengdu boasts a vast market in central and western China [17]. Chengdu Shuangliu International Airport has emerged as the fourth largest aviation hub in the country and the busiest international airport in central and western China [60]. As an international regional railway hub, Chengdu is committed to advancing the construction of outbound railway channels, aiming to establish a transport network that connects north to south and east to west [18]. Additionally, Chengdu is a key expressway hub in the west. The city plans to enhance its expressway network to ensure comprehensive coverage of major industrial areas and improve connectivity within the city, facilitating better north–south and east–west travel [28]. Moreover, Chengdu is set to develop a highway framework that will link western cities [59], forming an extensive network that covers the entire province, radiates throughout central and western regions, and integrates with the national highway system. The city's rapid development in technology and innovation sectors has made it a significant player in China's tech industry [13]. Additionally, Chengdu's universities and research institutions contribute to its reputation as an educational hub.

Chengdu displays a rich history and vibrant cultural heritage. Recognised as one of China's earliest famous historical and cultural cities, it is also listed among the top ten ancient capitals. With a continuous civilization spanning 4500 years, Chengdu is a national historical and cultural landmark. Remarkably, the city has retained its original site and name since its establishment. Expanding on its historical significance, Chengdu is home to numerous ancient relics and historical sites, such as the

A. Cheshmehzangi and T. Tang, *30 Years of Urban Change in China's 10 Core Cities*, Urban Sustainability, https://doi.org/10.1007/978-981-97-8846-0_7

Wuhou Shrine and the Jinsha Site Museum, which provide insights into its illustrious past [10]. Chengdu's traditional teahouses, opera performances, and festivals, such as Sichuan Opera and the Lantern Festival, vividly illustrate its extensive cultural heritage. This enduring cultural richness has established Chengdu as a major hub for cultural tourism and a symbol of China's deep historical narrative. Notably, Chengdu boasts the highest number of World Heritage projects in central and western China. The city is home to numerous historical sites and cultural landmarks, including the Dujiangyan Irrigation System, Wuhou Shrine, Du Fu Thatched Cottage, and the Jinsha Site Museum. These sites not only highlight Chengdu's historical significance but also contribute to its reputation as a city deeply rooted in ancient traditions and cultural wealth.

Chengdu serves as the scientific research hub of western China, featuring rich science and education resources alongside a vibrant research environment. The city is dedicated to establishing a nationally influential science and technology innovation hub. Chengdu Science City is envisioned as a nucleus for major scientific and technological infrastructure, hosting research institutes and university innovation platforms in the western region. It aims to develop a comprehensive national science center, focusing on areas such as cybersecurity, aerospace, and life sciences. This initiative positions Chengdu as a key player in frontier exploration and original innovation [49]. With robust support from the city, Chengdu Science City has rapidly emerged as the leading area for scientific and technological innovation in western China. It is renowned for its exceptional research environment, rapid accumulation of innovation resources, and high visibility. This development underscores Chengdu's commitment to advancing cutting-edge research and fostering an ecosystem conducive to technological breakthroughs and scientific excellence.

2 Chengdu's Physical Development Over the Past 30 years

- *City scale*

Over the past 30 years, Chengdu has undergone three comprehensive revisions of its urban master plan, continuously enhancing urban and rural planning and construction [35]. These revisions have focused on clarifying urban functional zoning and optimizing spatial layout, gradually transforming Chengdu into a regional central city characterized by diversified functions, complete facilities, robust comprehensive strength, and modern comprehensive service capabilities [4]. The 1996 version of Chengdu's urban master plan established a structure centered around the core of the central city, expanding development along six major traffic arteries. It emphasised adjusting the industrial structure in the eastern suburbs and proposed a development strategy extending towards the east and south [4]. In 2011, the urban master plan shifted the focus from a single-center development model to a multi-center development approach, emphasising integrated urban and rural planning [33, 58]. This plan aimed to balance development across various urban areas, fostering more cohesive growth [4]. By 2016, Chengdu advanced the differentiated development of functional spaces across its five regions: east, west, south, north, and central. This approach aimed to enhance the distinct characteristics and strengths of each area, promoting balanced and sustainable urban development [4]. The city's strategic planning efforts have ensured its evolution into a dynamic metropolitan hub with a well-coordinated urban framework, supporting its growth as a leading regional center (Figs. 1, 2, 3, 4 and 5).

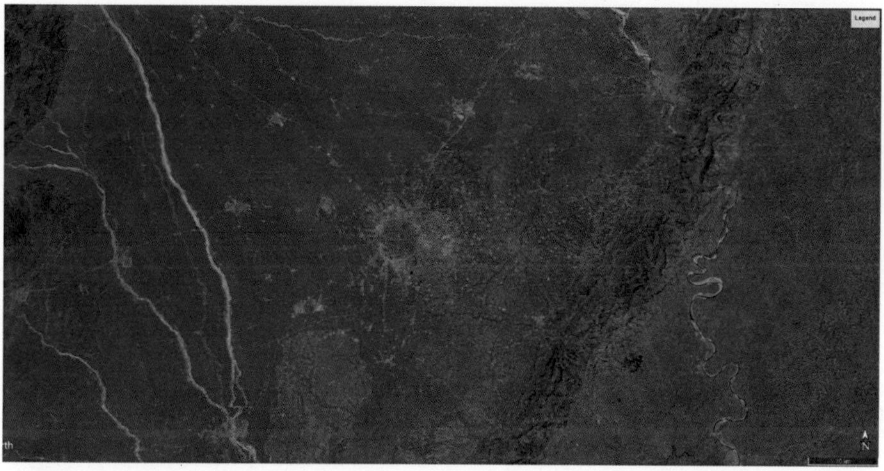

Fig. 1 Chengdu and its urban development areas in 1984 (map is extracted from open and accessible maps on Google Earth, generated in April 2024)

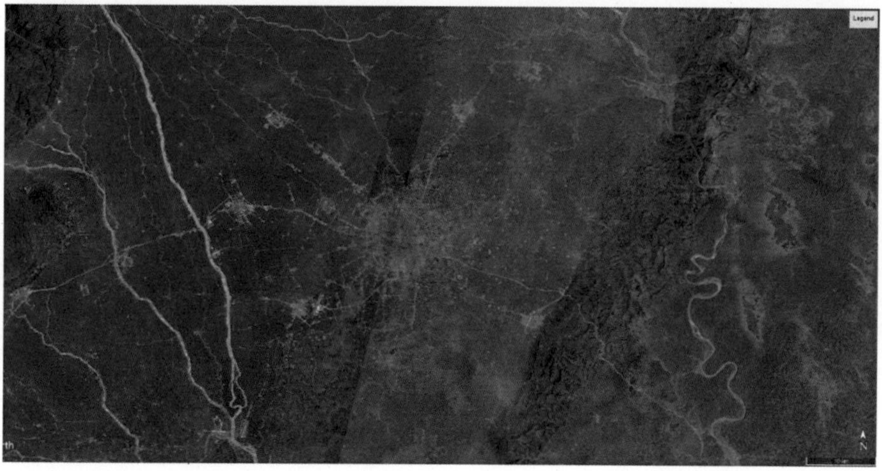

Fig. 2 Chengdu and its urban development areas in 1994 (map is extracted from open and accessible maps on Google Earth, generated in April 2024)

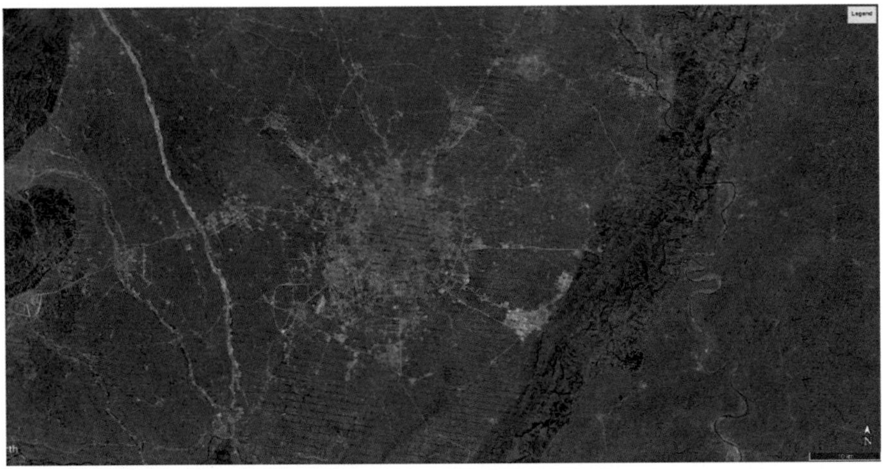

Fig. 3 Chengdu and its urban development areas in 2004 (map is extracted from open and accessible maps on Google Earth, generated in April 2024)

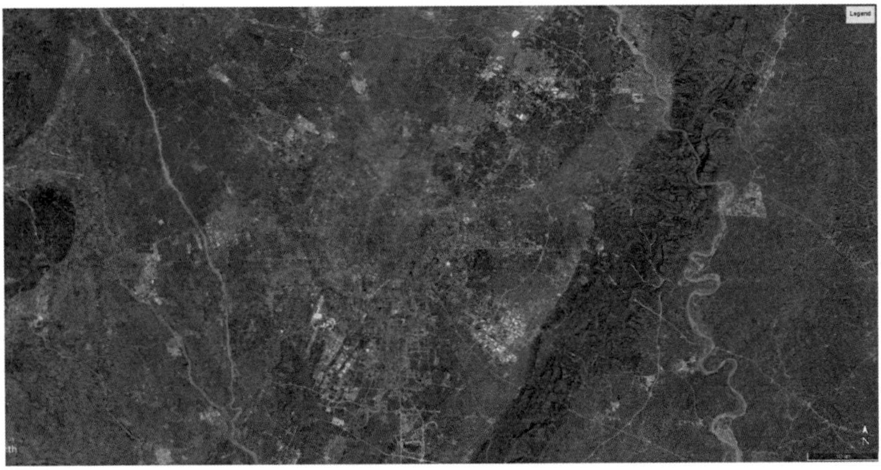

Fig. 4 Chengdu and its urban development areas in 2014 (map is extracted from open and accessible maps on Google Earth, generated in April 2024)

Chengdu's city-scale development has been marked by strategic initiatives aimed at enhancing its infrastructure, economic vitality, and liveability [57]. Major projects such as the Tianfu New Area have been pivotal in driving economic growth, attracting investments, and fostering high-tech industries. The establishment of extensive green belts and eco-friendly zones reflects the city's commitment to sustainable development, balancing urbanisation with environmental preservation [11]. Chengdu has also focused on improving public transportation networks, including expanding its

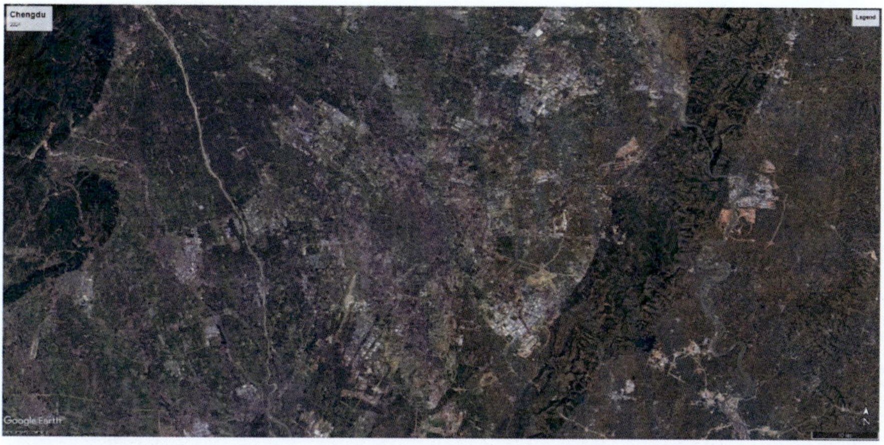

Fig. 5 Chengdu and its urban development areas in 2024 (map is extracted from open and accessible maps on Google Earth, generated in April 2024)

metro system and integrating smart city technologies [39] to enhance urban mobility [20]. Cultural and recreational facilities have been developed to boost residents' quality of life, making Chengdu not only a commercial and industrial powerhouse but also a vibrant and liveable city.

- *Residential Construction*

Chengdu has vigorously advanced residential housing construction with the objective of creating a highly liveable environment. By 2017, the per capita floor area of housing in Chengdu had risen dramatically from 9.7 square meters in 1997 to 42.38 square meters, with the average floor area per household reaching 110.6 square meters. These improvements signify a substantial enhancement in living conditions. The city's housing policies have incorporated scientific planning principles, focusing on sustainable urban development and efficient land use. Advanced building technologies and materials have been employed to ensure energy efficiency and environmental sustainability in residential construction [42]. Additionally, Chengdu has implemented comprehensive community planning, integrating green spaces, public amenities, and transportation networks to foster cohesive and vibrant neighbourhoods. These efforts reflect Chengdu's commitment to not only increasing housing availability but also enhancing the overall quality of urban living through scientifically informed development strategies.

In 1992, Chengdu initiated the comprehensive Funan River regulation project, encompassing river management, sewage interception, road construction, riverside greening, and the reconstruction of the old city [34]. This initiative significantly enhanced the living environment in Chengdu, catalysing a rapid increase in the development of modern residential areas. Beginning in 1996, Chengdu intensified its efforts to implement a strategic urban expansion towards the east and south. This strategic move was aimed at designating these areas as primary zones for future housing development, thereby significantly broadening the scope of real estate investment in the region [34]. Consequently, with ongoing urban development, there has been a harmonious alignment between housing construction and the overall urban layout in Chengdu.

- *Urban environment*

From 2002 onward, Chengdu has consistently ramped up its efforts in constructing environmental protection infrastructure projects. These endeavours, exemplified by the "six major projects," including the reconstruction of the old city, industrial zone structural adjustments in the eastern suburbs, the Smooth Traffic Project, water environment treatment, greening initiatives, and comprehensive urban environmental enhancements [34], have seen the full implementation of green construction methodologies, significantly enhancing Chengdu's urban landscape. Since the 1970s, urbanisation and the consequent expansion of Chengdu's city limits have led to varying degrees of pollution in its rivers. Following the comprehensive Funan River regulation project's completion in 1992, Chengdu embarked on an extensive urban water environment control project in 2002. This large-scale initiative primarily focused on renovating the sewage pipe network, resulting in a substantial improvement in the water quality of major urban rivers [50]. In addition, Chengdu have strictly implemented the prevention and control of air pollution, introduced several treatment plans, designated key areas for the prevention and control of air pollution, and set limits for the discharge of air pollutants. Chengdu has resolutely implemented the concept of green development and given greater prominence to ecological progress [2], and it was selected as one of the first prefabricated building demonstration cities in China [14–16], and successfully established a national ecological city and a national low-carbon pilot city.

3 Chengdu's Development Strategies

Here, we delve into three key development strategies of Chengdu, mainly focused on (1) Building a modern economic system, (2) promoting ecological progress aligned with the national agenda, and (3) cultural conservation and development. All three provide a critical overview of Chengdu's city-level and regional-level development patterns that have progressed more rapidly in recent years.

- *Building a modern economic system*

Chengdu has prioritised the advancement of the tangible economy and expedited the seamless amalgamation of internet technology, big data, and artificial intelligence with concrete economic activities. This emphasis primarily centers on fostering the growth of the digital economy, smart economy, green economy, creative economy, flow economy, and sharing economy. Efforts are directed towards establishing an interconnected and sustainable industrial ecosystem, fostering a modern economic framework characterised by three principal sectors: advanced manufacturing, modern

Fig. 6 Land made available for new industrial zone development in 1990s (map is extracted from open and accessible maps on Google Earth, generated in April 2024)

services, integrated industry, and contemporary urban agriculture. Chengdu is deeply engaging in international trade dynamics, actively embedding itself within the global industrial and value chains' core structures [9].

In its pursuit of a modern industrial system boasting global competitiveness and propelled by scientific research and innovation [4], Chengdu has concentrated efforts on nurturing scientific research zones such as Gaoxin Zone and Tianfu Science City [46]. These zones harness the foundational research and innovation prowess of universities and scientific institutions, fostering a comprehensive innovation ecosystem spanning basic research, technological development, and application to facilitate seamless knowledge transfer and transformation [4]. This strategic emphasis underscores Chengdu's commitment to driving sustained economic advancement through research-driven initiatives (See Figs. 6, 7 and 8).

- *Promoting ecological progress*

Situated in the upper reaches of the Yangtze River Basin, Chengdu serves as a pivotal conservation zone for water resources and biodiversity within China [24]. Encompassing diverse geomorphic features, ecosystems, land utilisation patterns, ecological safeguards, and territorial integrity, Chengdu exhibits a comprehensive representation of national ecological characteristics [52]. The establishment of Chengdu as an ecological civilization pilot demonstration area assumes significance in delineating a scientific developmental trajectory for regions dedicated to ecological preservation [4]. This initiative underscores Chengdu's pivotal role in advancing ecological stewardship on a national scale, serving as a beacon for sustainable development practices in ecologically sensitive regions [6]. Aligned with national ecological development plans, Chengdu's strategic positioning as a pioneer in ecological civilization underscores its commitment to broader environmental agendas

Fig. 7 South-east expansion of Chengdu with regional industrial development plans in early 2000s to 2010s (map is extracted from open and accessible maps on Google Earth, generated in April 2024)

Fig. 8 New industrial zone with regional capacity and influence as part of new urban development zones of Chengdu as it stands in 2024 (map is extracted from open and accessible maps on Google Earth, generated in April 2024)

[55]. By integrating sustainable practices into urban planning, Chengdu exemplifies China's vision for ecological progress. Through initiatives such as afforestation programs, wetland conservation efforts, and sustainable resource management, Chengdu actively contributes to the realisation of national ecological goals. Its role

as a pilot demonstration area signifies a proactive approach towards fostering ecological resilience and promoting harmonious coexistence between humans and nature [30], in accordance with national directives.

In a concerted effort to foster synergy between ecological and urban spaces, enhance the quality of residential environments, and uphold ecological integrity, Chengdu has established a demonstration area dedicated to biodiversity preservation and the creation of the Longquanshan Urban Forest Park, thus cultivating a premier urban green hub. Embracing the ethos of ecological civilization, Chengdu is committed to strengthening the ecological protection and conservation functions of Longquan Mountain, intensifying efforts to restore and preserve its ecological balance [4]. Measures include safeguarding rivers, lakes, and wetlands, enhancing forest protection and utilisation, rigorously safeguarding prime agricultural land, and elevating overall ecological standards. Chengdu is furthering environmental enhancement endeavours by implementing comprehensive measures to curtail tailpipe emissions, mitigate dust proliferation, and prohibit open-air burning. Concurrently, it is bolstering initiatives to combat water pollution, safeguard aquatic ecosystems, and deploying targeted strategies to mitigate pollutant discharge and forestall environmental hazards [4], thereby remedying deficiencies in ecological and environmental quality.

- Cultural conservation and development

As one of China's earliest designated historical and cultural cities, Chengdu hosts a wealth of archaeological treasures and cultural landmarks [32]. Its systematic approach to heritage conservation incorporates advanced methods such as architectural restoration techniques and archaeological excavations [26, 40]. By safeguarding historical buildings, environments, and intangible cultural heritage, Chengdu not only preserves its rich past but also fosters a deeper understanding of its cultural evolution across different historical epochs [37]. Therefore, Chengdu, renowned for its historical and cultural significance, employs cutting-edge techniques [4] in archaeological research and architectural preservation to safeguard its heritage, providing insights into its cultural evolution over time.

The advancement of cultural and creative industries serves as a cornerstone in Chengdu's endeavour to enrich both local and global cultural dialogues [4]. Emphasising the synergy between cultural and creative sectors, Chengdu integrates these domains with science, technology, tourism, ecology, and allied industries. This holistic approach underscores Chengdu's commitment to fostering a multifaceted urban landscape. Moreover, Chengdu actively nurtures its traditional culture and art, harnessing their inherent local flavours [4]. Efforts are directed towards enhancing the creative design of tourist souvenirs, encapsulating the essence of folk traditions, historical narratives, and geographical uniqueness, thereby contributing to the branding of Chinese tourism commodities. Furthermore, Chengdu is dedicated to enhancing tourism infrastructure and public amenities, envisioning the city as a quintessential cultural and characteristic model, with tourism [4] serving as the conduit for its cultural dissemination and enhancement (See Figs. , 10, 11 and 12).
9

Fig. 9 Urban expansion in southern regions from 1990s to 2024. Map is from early 1990s. (map is extracted from open and accessible maps on Google Earth, generated in April 2024)

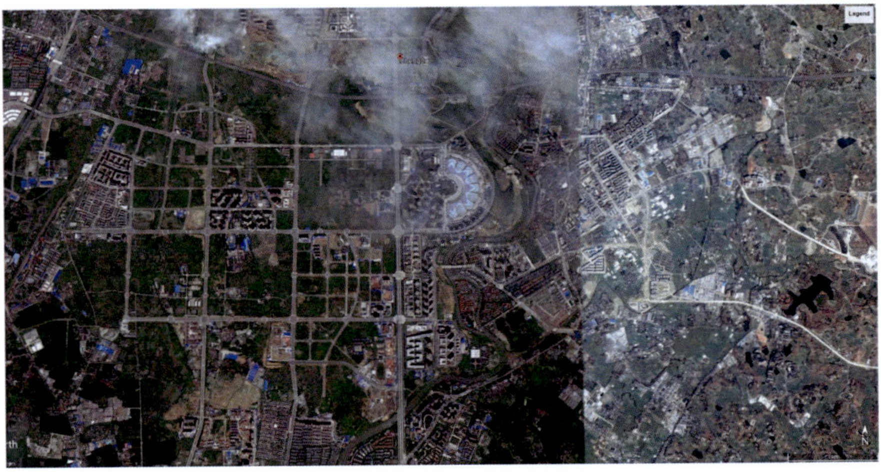

Fig. 10 Urban expansion in southern regions from 1990s to 2024. Map is from 2000s. (map is extracted from open and accessible maps on Google Earth, generated in April 2024)

3.1 *Chengdu Science City: An Exemplar for Talent Cultivation and High-Quality Ecological Development*

Chengdu Science City (CSC), located in Tianfu New District, Chengdu, is a new modern city integrating government, enterprises, universities, research institutes and end users. It has a planned area of 125 square kilometers and a core area of

Fig. 11 Urban expansion in southern regions from 1990s to 2024. Map is from mid 2010s. (map is extracted from open and accessible maps on Google Earth, generated in April 2024)

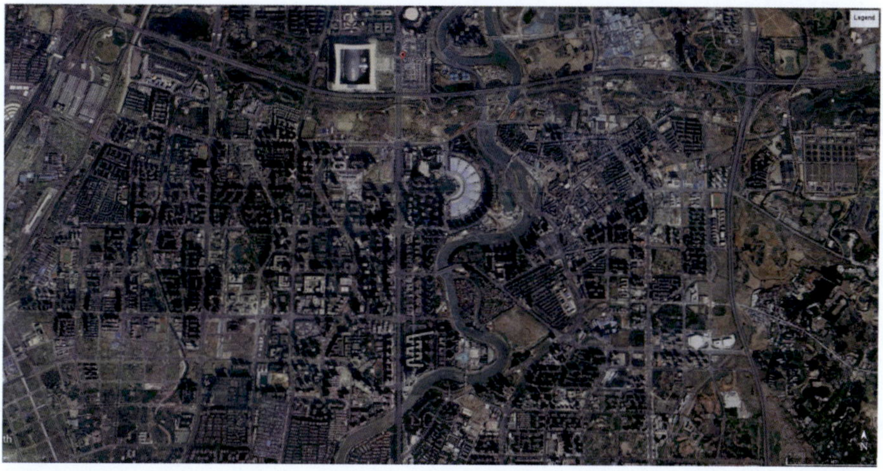

Fig. 12 Urban expansion in southern regions from 1990s to 2024. Map is from 2024. (map is extracted from open and accessible maps on Google Earth, generated in April 2024)

73 square kilometers [43]. Chengdu Science City has established a comprehensive innovation ecosystem, spanning from fundamental scientific research, through to application development, and eventual industrial fruition. Encircling the Xinglong Lake area, it hosts six distinct industrial clusters, encompassing scientific research and application conversion, information networks, biomedicine, creative design, military-civilian integration, and Sino-German collaborations [44]. Primarily oriented towards cultivating high-value sectors such as fundamental research, internet

technologies, biotechnology, advanced manufacturing, modern finance, and creative design, Chengdu Science City aims to evolve into a pivotal hub for technological innovation and scientific advancement [44]. It aspires to be recognized as a beacon of innovation and a focal point within the comprehensive national science landscape.

Chengdu Science City adopts a strategic approach focused on attracting high-caliber talent and enterprises by emphasizing the provision of superior ecological surroundings, urban amenities, and comprehensive facilities. Its development blueprint entails the systematic advancement of key infrastructural components, public amenities, industrial infrastructure, residential accommodations for talent, and ancillary support services [49]. The establishment prioritizes the cultivation of a favourable ecological milieu, exemplified through initiatives like the Luxi Wetland, Zhigu Greenway, Fengqi Wetland, with ecological spaces comprising 42% of the built-up area. Furthermore, Chengdu Science City seeks to foster synergies with Chongqing Science City, particularly in technology transfer, talent exchange, research funding, thereby fostering reciprocal collaboration and mutual benefit. Strengthening the collaboration between Chengdu and Chongqing aims to nurture a regional innovation ecosystem grounded in cooperative endeavours [49].

3.2 Chengdu Tianfu International Airport

Upon its inauguration, the construction of Chengdu Tianfu International Airport positioned Chengdu as the third city in China to possess two commercial/passenger airports. This recent infrastructure augmentation significantly catalysed Chengdu's evolution into an international aviation hub, augmenting the city's global connectivity and unleashing its developmental prospects. Recognising the imperative for a Western Chinese city to possess aerial dominance to spearhead regional integration into the global economy, Chengdu's dual-airport configuration assumes paramount importance. Consequently, Chengdu emerges as the sole major city in Western China endowed with air superiority, a status poised to further bolster the region's international prominence and connectivity [48].

Located southwest of Jianyang City, Shuangliu International Airport is strategically positioned to complement the emerging developmental trajectory of Chengdu towards the east, thereby fostering the growth of the Chengdu-Chongqing twin cities economic circle. This synergy is pivotal in driving the collective advancement of Chengdu, Deyang, Meishan, Ziyang, and other neighbouring urban centers. Each airport in Chengdu's aviation landscape harbours distinct strategic objectives. Tianfu International Airport aspires to evolve into a premier international passenger and cargo aviation hub, composing a transit nexus with Chengdu at its nucleus, spanning key global destinations across Europe, North America, Australia, New Zealand, Southeast Asia, and Northeast Asia. Meanwhile, Shuangliu International Airport primarily services domestic and regional routes, focusing on strengthening connectivity within mainland China and neighbouring regions. This delineation of roles facilitates specialized development trajectories for each airport, while also

Fig. 13 Development of Chengdu Tianfu International Airport from an empty land to a major regional airport. Map is from early 2000s. (map is extracted from open and accessible maps on Google Earth, generated in April 2024)

fostering interdependence and complementarity between them. The potential for mutual reinforcement and synergy between the two airports enables them to function as alternate facilities for one another, culminating in the gradual establishment of a dual-hub airport system characterised by coordinated growth, complementary functionalities, and shared developmental objectives. This synergistic evolution reflects Chengdu's strategic vision for cultivating a robust and resilient aviation infrastructure [31] to meet the diverse demands of its expanding regional and global connectivity aspirations (Figs. 13, 14 and 15).

3.3 Eastern Suburb Memory

The government actively encourages the advancement of emerging service sectors, such as cultural creativity and design services, fostering synergies with allied industries. Cultural entrepreneurship emerges as a focal point in urban development, aligning with sustainable urban strategies, notably emphasising green and low-carbon initiatives [19]. By integrating cultural industries with tourism, there emerges the opportunity to enhance the tourism value chain, facilitating the sustainable development of resources, culture, ecology, economy, and society.

Eastern Suburb Memory stands as an iconic cultural and creative park in Chengdu, serving as the sole urban industrial land renewal and industrial heritage conservation initiative in Sichuan Province. Recognised as a national hub for the music industry and a fundamental commercial district within China, it emerges as a noteworthy landmark for Chinese tourism and cultural endeavours. Furthermore, it serves as a

Fig. 14 Development of Chengdu Tianfu International Airport from an empty land to a major regional airport. Map is from early 2010s. (map is extracted from open and accessible maps on Google Earth, generated in April 2024)

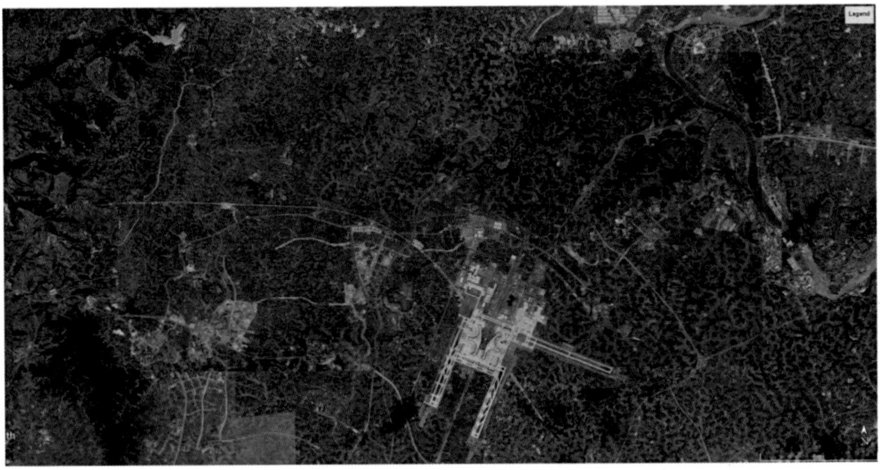

Fig. 15 Development of Chengdu Tianfu International Airport from an empty land to a major regional airport. Map is from 2020s. (map is extracted from open and accessible maps on Google Earth, generated in April 2024)

model demonstration site for the cultural industry in Sichuan Province, seamlessly blending creative, cultural, and tourism sectors within its innovative and cultural landscape.

Situated on the grounds of a disused industrial factory, Eastern Suburb Memory undergoes rejuvenation through a unique industrial preservation approach, harmonizing the conservation of industrial heritage with cultural and entrepreneurial pursuits. This transformation seamlessly integrates the industrial aesthetics emblematic of the planned economy era with the contemporary functionalities of modern commercial structures [56]. The site juxtaposes architectural relics from the 1950s Soviet era with state-of-the-art office complexes, encapsulating the evolution of industrial landscapes across different historical epochs within a single locale [38]. Eastern Suburb Memory serves as a custodian of Chengdu's industrial legacy while concurrently evolving into a vibrant nexus for diverse cultural expressions spanning music, visual arts, theatre, photography, and more. Beyond nurturing its own industrial ecosystem, it exemplifies a pioneering model for media-driven urban development. Presently, hosting close to 100 enterprises, it generates an annual output nearing 100 million yuan [56], solidifying its status as a contemporary, globally-oriented center for cultural and creative industries in Chengdu.

4 Chengdu's Local Governmental Projection for Future

- ***Spatial pattern optimisation of urban development***

Chengdu prioritises the impact of urban master planning on its developmental trajectory, meticulously executing territorial and spatial planning initiatives to expedite the realisation of a city aligned with contemporary development paradigms [4]. This entails the cultivation of Chengdu into a national center city, characterised by its status as a beautifully liveable park city, an international metropolis, and a global cultural hub, all underpinned by a commitment to sustainable growth. In broadening the scope of urban development and refining the distribution of urban functions, Chengdu advocates for a regionalised approach to development strategy. By accounting for the inherent environmental capacity of resources and conducting assessments on the suitability of land and spatial development [4], the city endeavours to foster differentiated development across its five regions: east, west, south, north, and central. Through this nuanced approach, Chengdu aims to optimize the utilisation of resources and enhance the resilience of its urban fabric [4].

The eastern sector places emphasis on advancing the establishment of the Chengdu-Chongqing twin-city economic circle, proactively facilitating the integrated growth of the Chengdu-Chongqing urban cluster. This involves fostering the development of Konggang New City, Jianzhou New City, Huaizhou New City, and Jianyang City through a coordinated and clustered urban layout approach [4]. This initiative offers an expanded arena for Chengdu's development, accommodating the relocation of populations from the central urban core and alleviating environmental strain therein. Concurrently, it reinforces the ecological safeguarding of Longquan Mountain Urban Forest Park, while strategically charting the establishment of a modern new city [47] geared towards advanced manufacturing and service industry bases.

The western area features favourable ecological attributes, characterised by abundant light, heat, and water resources, rendering it highly conducive to agricultural activities [8]. Taking into account the ecological characteristics of the western region, the plan is to leverage this area for the advancement of green industries such as biomedicine, organic agriculture, cultural innovation, and afforestation. It aims to establish this zone as Chengdu's paramount ecological hub, a vital region for vegetable and grain production, a demonstration zone for green and low-carbon technology industries, and the nucleus of a globally renowned tourist destination [4]. At the same time, the city implement the requirements for ecological protection in the upper reaches of the Yangtze River Economic Zone, and protects important water sources and the Dujiangyan irrigation area.

In the southern region, the emphasis lies on fostering innovative development paradigms. Tianfu New Area is poised to serve as the nucleus for establishing a showcase area for high-quality development and a national demonstration zone for independent innovation. Central to this endeavour is the cultivation of innovation and creativity, coupled with the consolidation of emerging functions. Additionally, there will be a moderate expansion of construction land to implement a growth model that places equal emphasis on development and sustainability [53]. This strategy aims to propel the southern region towards the forefront of innovative economic and technological advancement, positioning it as a catalyst for Chengdu's overall growth trajectory [8].

Simultaneously, the northern sector prioritises the execution of the Belt and Road initiative and the blueprint for the western land-sea corridor. Leveraging the Chengdu International Railway Port, initiatives include establishing the Chengdu-Europe Railway Port Economic and Technological Development Zone and the Chengdu International Railway Port Comprehensive Bonded Zone. These efforts aim to forge a cross-regional high-tech industrial belt and an exemplary region for military-civilian integration. Ultimately, the objective is to construct an industrial framework aligned with the imperatives of an open economy [4]. Additionally, the central region struggles with several existing challenges, including high development intensity, large-scale construction, low industrial sophistication, inadequate public amenities, and subpar urban quality. Consequently, the central region's enhancement strategy focuses on continual refinement of the urban spatial layout and hastening organic urban revitalisation. This involves proactive development of urban and modern service industries to foster a vibrant, high-energy urban environment conducive to high-quality living standards [4].

- ***Break through regional limitations and promote intra-city development***

Addressing the issue of regional development disparity stands as a key challenge in Sichuan Province's progression. Moving forward, urban integration transcends individual city competition to entail the clustering of city circles [25]. The overarching mandate is for the state to fervently advocate for urban agglomerations, fostering collaboration to drive regional advancement collectively. The 19th National Congress of the Communist Party of China highlights the imperative of implementing a strategy for regional coordinated development, prioritising the harmonised growth of cities within urban agglomerations [1]. For Chengdu, this necessitates transcending administrative boundaries and fostering coordinated development with surrounding cities and prefectures such as Deyang, Mianyang, Ziyang, Meishan, Leshan, Ya'an, among others. Given their geographical proximity, extensive human interactions, and economic interdependencies, these cities possess a solid foundation for development. Additionally, they boast advantages in natural ecology, industrial infrastructure, and innovative dynamism [3]. Ecological preservation and sustainable development are deemed essential prerequisites for urban growth, with collaborative efforts directed towards comprehensive protection of the Yangtze River's ecological milieu through regional safety layout planning. Concurrently, enhancing the contribution of technological innovation zones to development has emerged as a pivotal strategy within the Chengdu metropolitan area [3], encompassing the strategic delineation of regional industrial functional zones to boost high-quality industrial advancement.

In summary, Chengdu's developmental trajectory is underpinned by its robust industrial and economic groundwork. This progress is not only attributable to Chengdu's adept leveraging of significant opportunities, including the Third Front movement, the reform and opening-up policy, the western region's development, and the Belt and Road initiative (BRI) [21], but also to its proactive pursuit of these opportunities [4]. Looking ahead, Chengdu's strategic blueprint entails further optimisation of its urban spatial configuration, alongside the implementation of pivotal livelihood initiatives. Additionally, the city will concentrate its efforts on fostering advancements in biotechnology, intelligent manufacturing, and other strategic sectors, aiming for significant breakthroughs. These endeavours are geared towards enabling Chengdu to make substantive contributions to the continued development of the western region [4] and the realisation of the Belt and Road initiative's objectives.

5 Chengdu's Challenges of the Past and the Present

The first overarching challenge of Chengdu is its over-population and diverse impacts as a result of such phenomena. As central cities rapidly evolve, the issue of over-population in these urban hubs is becoming more pronounced [8]. The basis of a region's economic growth rests on the synchronised advancement of industry and urban infrastructure. Overcrowding in cities results in a shortage of space necessary for industrial expansion and places significant strain on sustainable urban growth and public services. Consequently, the economic development of these areas is adversely affected [8]. Besides, it also affects the improvement of the quality and standard of urban life. To mitigate these challenges, it is essential to promote balanced regional development by enhancing infrastructure in suburban and rural areas [7], thereby alleviating the pressure on central cities and fostering a more equitable distribution of population and resources. Additionally, implementing smart city planning and sustainable practices can further support long-term economic stability and growth.

To address these constraints, Chengdu intends to shift its development paradigm from a focus solely on the urban core to a comprehensive strategy that includes both the central city and its suburban areas. By expanding development into the suburbs,

Chengdu aims to create additional space for growth, thereby alleviating issues related to population density, housing shortages, and traffic congestion in the city's central regions [8]. The urbanisation of Chengdu's suburbs is advancing systematically, underpinned by robust infrastructure and effective public management. These foundational elements are fostering a conducive environment for integrated development, thereby enhancing the city's overall capacity for sustainable growth [27]. Furthermore, this strategic approach promotes a balanced regional development model that leverages suburban potential, ensuring equitable distribution of resources and opportunities, and contributing to the long-term economic and social vitality of Chengdu. The success of this model can serve as a blueprint for other rapidly growing cities facing similar urbanisation challenges.

Chengdu's second foremost challenge is its geographical limitations. Located in the southwestern inland region, Chengdu faces significant developmental constraints due to its distance from coastal and border ports [29]. This relative isolation poses considerable challenges, including extended distances for external transportation, elevated logistics costs, and limited ease of movement for people. Such factors impede Chengdu's ability to access global markets, participate in international competition, and integrate into the global economy [36]. The geographical remoteness hinders the city's economic dynamism by creating logistical bottlenecks that elevate the cost of goods and services, ultimately reducing its competitive edge on the global stage. Moreover, the difficulty in connecting with international partners and markets stifles opportunities for economic expansion and innovation, essential for thriving in a globalised economy [8]. To overcome these barriers, Chengdu needs to invest in improving transportation infrastructure, enhancing connectivity with other economic hubs, and developing strategic partnerships that can help mitigate the disadvantages of its inland location [59]. By doing so, Chengdu can better position itself as a significant player in the global economy, fostering sustainable growth and development.

Since the implementation of the reform and opening-up policy, Sichuan province has been dedicated to expanding its openness to the outside world and overcoming the developmental challenges faced by inland regions. The province has proactively seized strategic opportunities presented by China's Belt and Road Initiative (BRI) to establish a comprehensive framework for external engagement [41]. By capitalising on the advancements in modern aviation and high-speed rail networks, along with breakthroughs in modern information technology, Sichuan has significantly enhanced its economic growth. This strategic approach has enabled the province to improve its connectivity with global markets, facilitating increased trade and investment. The development of robust transportation infrastructure and the integration of cutting-edge technologies have been pivotal in transforming Sichuan into a dynamic economic hub. Furthermore, these efforts have positioned the province as a crucial participant in the Belt and Road Initiative, fostering stronger economic ties and collaborative ventures with other regions. As a result, Sichuan's economy has experienced substantial growth, driven by improved accessibility, innovation, and a more open economic environment [23].

In this new era of development, Chengdu must transcend the geographical constraints of its basin location to achieve significant economic advancement in the inland region. In 2017, Chengdu introduced the ambitious "East Advance" strategy, which includes a comprehensive development plan for a new urban area on the eastern side of Longquan Mountain [45]. This strategy is designed to disrupt the existing urban development pattern, thereby expanding the spatial boundaries for the city's growth. By shifting the focus eastward, the "East Advance" initiative aims to alleviate congestion in the central areas, distribute development more evenly, and provide new opportunities for economic activities and infrastructure development [45]. This expansion is expected to foster a higher level of economic growth, enhance regional connectivity, and attract investments, positioning Chengdu as a more dynamic and competitive inland city. The strategic development of the eastern region not only opens up new spaces for urban construction but also paves the way for sustainable and balanced urban expansion, ensuring Chengdu's continued prosperity and integration into broader economic networks [45]. This approach leverages key transportation hubs like Tianfu International Airport, the Chengdu-Europe Express Railway, the Chengdu-Chongqing High-speed Railway, and the Yangtze River Waterway to spur the coordinated development of nearby urban areas and facilitate integration into the Belt and Road Initiative.

6 Lessons Learnt from Chengdu for Other Cities or Contexts

A major lesson learnt from Chengdu is its plan and direction for park city construction, boosting an opportunity for green and ecological development patterns. Urbanisation brings economic growth and city expansion, but improper resource allocation can result in issues like population surge, traffic congestion, housing shortages, and environmental pollution [12]. To address these challenges, Chengdu initiated the construction of a park city as a pilot demonstration area. This approach adheres to urban development principles by scientifically arranging spaces for production, living, and ecology, and integrating ecological, cultural, and economic values into urban planning. By developing green industries, providing green products, and promoting a green economy, Chengdu aims to transform ecological values into drivers of development [51]. Chengdu has undertaken several major projects that have resulted in the addition of 19 million square meters of green space, increasing the city's forest coverage rate to 40.2%, which is 1.9% points higher than it was five years ago. The green coverage rate in developed areas is now 43.5%, with the per capita green area in parks reaching 14.9 square meters. In terms of enhancing urban green spaces, the Tianfu Greenway system has expanded to a total of 4408 km, and 1018 community greenways have been established [5].

In exploring the approaches and methods for park city construction, Chengdu has innovated its urban management concepts and practices. The city has advanced comprehensive area development and the creation of future communities, ensuring natural and orderly urban growth. These efforts have yielded numerous advanced experiences and practices that contribute to the new development concept of city construction [51]. Other lessons learnt from Chengdu that can benefit other cities or contexts include: first, the importance of integrated urban planning, where Chengdu's approach to park city construction demonstrates the effectiveness of harmonising economic development with environmental sustainability and quality of life improvements. Second, Chengdu's emphasis on green infrastructure development [22] highlights the value of investing in parks, greenways, and forest coverage to mitigate urban challenges such as population growth, traffic congestion, and pollution, while enhancing residents' well-being. Lastly, Chengdu's commitment to innovation in urban governance and community engagement underscores the significance of adaptable governance structures [54] and citizen participation in shaping resilient and vibrant cities.

References

1. Basso M, Wang L (2022) China's national urban agenda: Transition and reframing cities. Identifying models of national urban agendas: a view to the global transition. Springer International Publishing, Cham, pp 77–100
2. Bum T (2024) "Doing the work of science": Tibetan environmentalists and the changing dynamics of ecological conservation in China. China Q: 1–16
3. CDCPPCC.gov.cn (2021) How can the four cities join hands to play as" first move" for national strategy? 四城携手 如何下好国家战略"先手棋"?>. Available from: https://www.cdcppcc.gov.cn/show-1023-110039644-1.html
4. CDDRC.chengdu.gov.cn (2018) Outline of the 13th five-year plan for national economic and social development of Chengdu. 成都市国民经济和社会发展第十三个五年规划纲要
5. CDDRC.chengdu.gov.cn (2021) The forest coverage rate of Chengdu reached 40.2%, and the amount of forest added in five years is equal to 3,090 tianfu squares. 成都森林覆盖率达40.2% 五年新增森林等于3090个天府广场>. Available from: http://cddrc.chengdu.gov.cn/cdfgw/fzggdt/2021-08/06/content_e7a6d71aa01e406aa48ac70e2b2b53d5.shtml
6. CECRPA (2014) System innovation to create ecological civilization of Chengdu sample. 制度创新打造生态文明的成都样本>. Available from: http://www.cecrpa.org.cn/sfcj/sdsf/201409/t20140929_637401.shtml
7. Chen JC (2014) "The green leap forward": environmentalization and rural-urban transformation in contemporary China. University of California, Berkeley. Available from: https://search.proquest.com/openview/f1694d75ca6043230398461c184b3626/1?pq-origsite=gscholar&cbl=18750
8. Chengdu Municipal Bureau of Planning and Natural Resources (2020) Differentiated development plan for five functional areas in Chengdu. 成都市五大功能区差异化发展规划>. Available from: http://www.planning.cn/2019/view.php?id=769
9. Chengdu New Economic Development Commission (2019) Chengdu's practice of developing new economy. 发展新经济的成都实践>. Available from: http://cdxjj.chengdu.gov.cn/xjjfzw/fzxjjdcdsj/2019-04/24/content_db615b6fcef445cca2c16693f20b54c5.shtml
10. Chengdu Wenming (2021) The Tianfu culture and humanistic of Chengdu. "天府文化"涵养人文成都>. Available from: http://cd.wenming.cn/wmbb/201707/t20170705_4591825.shtml
11. Cheshmehzangi A (2016) China's New-Type Urbanisation Plan (NUP) and the foreseeing challenges for decarbonization of cities: a review. Energy Procedia 104:146–152

12. Cheshmehzangi A (2020). Towards urban energy management: a brief review of China's recent transition in urban energy demand. Low Carbon Econ 11(01):1. Available from: https://www.scirp.org/html/1-2900311_102925.htm
13. Cheshmehzangi A (2023) Mapping urban regeneration: introduction to the new, the old, and the new-old. In: Mapping urban regeneration: city life experiences in Yunnan, China. Singapore, Springer Nature Singapore, pp 1–21
14. Cheshmehzangi A, Chen H (2021a) Key suggestions and steps ahead for China's carbon neutrality plan. China's Sustainability Transitions: Low Carbon and Climate-Resilient Plan for Carbon Neutral 2060. Singapore, Springer Nature, pp 189–206
15. Cheshmehzangi A, Chen H (2021b) Decarbonised race and new destination in China. China's sustainability transitions: low carbon and climate-resilient plan for carbon neutral 2060. Singapore, Springer Nature, pp 53–69
16. Cheshmehzangi A, Dawodu A, Sharifi A (2021a) Sustainable urbanism in China. Routledge, New York
17. Cheshmehzangi A, Tang T (2022a) China's city cluster development in the race to carbon neutrality. Singapore, Springer Nature.
18. Cheshmehzangi A, Tang T (2022b) Chengdu-Chongqing city cluster: from an economic zone to a major economic circle in the Western China. In: China's city cluster development in the race to carbon neutrality, Singapore, Springer Nature, pp 77–92
19. Cheshmehzangi A, Tang T (2022c) Carbon neutrality and China's city cluster development: Future directions. In: China's City Cluster Development in the Race to Carbon Neutrality. Singapore, Springer Nature Singapore, pp 189–199
20. Cheshmehzangi A, Tang T, Li S, Su Z (2022) Transportation and mobility as the number two enemy to the prevalence of the COVID-19 pandemic. J Infect Dis & Prev Med 9:246
21. Cheshmehzangi A, Xie L, Tan-Mullins M (2021b) Pioneering a green Belt and Road Initiative (BRI) alignment between China and other members: mapping BRI's sustainability plan. Blue-Green Syst 3(1):49–61
22. Cheshmehzangi A, Butters C, Xie L, Dawodu A (2021c) Green infrastructures for urban sustainability: issues, implications, and solutions for underdeveloped areas. Urban for & Urban Green 59:127028
23. China development observation (2019) From western Inland city to International emerging City: Chengdu's "Millennium Transformation" 从西部内陆城市到国际新兴城市:成都“千年之变”的世界雄心>. Available from: http://www.chinado.cn/?p=8220
24. Dong X, Yuan Q, Kou Y, Li S, Ren P (2023) Distribution and ecological network construction of national natural protected areas in the upper reaches of Yangtze river. Sustainability 15(2):1012
25. Fang C, Yu D, Fang C, Yu D (2016). The developmental strategies and basic principles for China's new urbanization. In: China's new urbanization: developmental paths, blueprints and patterns, pp 49–109
26. Guo N (2022) A model approach for conservation of historic blocks to address property rights and financing issues of historic cities in China. Available from: https://theses.lib.polyu.edu.hk/handle/200/11672
27. Jiang FF, Zhang JP (2019) Actively "break through the siege" speed up the construction of a comprehensive embodiment of the new development concept of the city. 积极“突围” 加快建设全面体现新发展理念的城市>. Available from: http://qstheory.cn/llqikan/2019-04/30/c_1124436338.htm
28. Jiang Q, Gan Y, Deng X (2024) Research on the development path of international supply chain hub city based on the manufacturing industry in Chengdu. Journal of Global Humanities and Social Sciences 5(4):162–167
29. Kettunen E (2019) The one belt one road initiative and the changing multi-scalar governance of trade in China. In: Mapping china's 'one belt one road'initiative, pp 117–141
30. Li L, Cheshmehzangi A, Chan FKS, Ives CD (2021) Mapping the research landscape of nature-based solutions in urbanism. Sustainability 13(7):3876
31. Liu WY, Wang JY (2021) Why Chengdu becomes the third city has two airports after Beijing and Shanghai. 京沪之后中国第三个拥有双机场的城市,为什么是成都?>. Available from: https://new.qq.com/omn/20210627/20210627A06VQF00.html

32. Liu Z (2023) Regeneration of Liangzhu culture: multimedia exhibition, simulated restoration, innovative cultural products, nearby area integration, virtual reality and augmented reality. HumIties Soc Sci Commun 10(1):1–18
33. Liu Z, Wang Y, Zhang C, Liu D (2023) Quantitative analysis of spatial heterogeneity and driving forces of the urban spatial structure's development level based on multi-source big data: a case study of Beijing. China Land 12(6):1178
34. Liu ZG (2018) The improvement of Chengdu's urban construction after the reform and opening-up policy. 改革开放以来日臻完善的成都城市建设>. Available from: http://www.scdfz.org.cn/bsszqk/bssz2021n
35. Miao JT (2019) Planning particularities: reinterpreting urban planning in China with the case of Chengdu. Plan Theory Pract 20(4):512–536
36. Qin X, Qian Y, Zeng J, Wei X (2022) Accessibility and economic connections between cities of the new western land–sea corridor in China—enlightenments to the passageway strategy of Gansu province. Sustainability 14(8):4445
37. Sichuan Government (2018) The "Chengdu model" for the protection of large sites is leading in China: the complete protection of urban culture. 大遗址保护"成都模式"在全国领先:完整保护城市文脉>. Available from: https://sichuan.scol.com.cn/cddt/201812/56755513.html
38. SINA News (2016) Characteristic Industrial park -- Eastern suburb memory. 特色产业园区——东郊记忆>. Available from: http://collection.sina.com.cn/whcy/2016-11-23/doc-ifxxwrwh5004185.shtml
39. Tan-Mullins M, Cheshmehzangi A, Chien S, Xie L (2017) Smart-eco cities in China: trends and city profiles 2016. Exeter, University of Exeter (SMART-ECO Project)
40. Teng F (2017) An effective framework of public-private partnerships: creating an economically viable plan for conserving and managing Beijing's urban heritage sites. Doctoral dissertation, Columbia University
41. Teo HC, Lechner AM, Walton GW, Chan FKS, Cheshmehzangi A, Tan-Mullins M, Chan HK, Sternberg T, Campos-Arceiz A (2019) Environmental impacts of infrastructure development under the belt and road initiative. Environments 6(6):72
42. The Chengdu Government (2017) Chengdu real estate Development five-year plan. 成都市房地产业发展五年规划>. Available from: http://gk.chengdu.gov.cn/govInfoPub/detail.action?id=1810989&tn=2
43. The People's Government of Sichuan Province (2021) Plate-unveiling ceremony of western China (Chengdu) science city and Tianfu laboratory held in Chengdu>. Available from: https://www.sc.gov.cn/10462/10758/10760/10765/2021/6/9/4dae426a1da7441e8d5aa2a2139e412a.shtml
44. The People's Government of Sichuan Province (2016). The accelerated rise of Chengdu Science City promotes the construction of national central city. 成都科学城加速崛起 助推国家中心城市建设>. Available from: https://www.sc.gov.cn/10462/10464/10465/10595/2016/10/31/10402105.shtml
45. Urban Planning Society of China (2017). Chengdu's new urban development strategy: To develop industries eastward and learn from Xiongan New Area. 成都城市发展新战略: 产业东进,学习雄安新区>. Available from: http://www.planning.org.cn/news/view?id=6709&cid=12
46. Walcott SM (2021) Science parks and high-tech zones. In: Fu X, Mckern B, Chen J (eds) The Oxford Handbook of China Innovation. Oxford, Oxford University Press, p 337
47. Wenjiang (2020) Focus on the key words of development and sing "A Tale of Two Cities" 读懂发展关键词 唱好"双城记">. Available from: http://www.wenjiang.gov.cn/wjzzw/zzssxx/2020-12/29/content_87a6c097f5de4faf928d61ea001838e4.shtml
48. www.nbd.com.cn (2020) Chengdu Shuangliu: exploring a new development model of dual airport cities in China. 成都双流:探寻国内双机场城市发展新模式>. Available from: http://www.nbd.com.cn/articles/2020-11-30/1560104.html
49. www.cdtf.gov.cn (2021) Chengdu science city. 成都科学城>. Available from: http://www.cdtf.gov.cn/cdtfxq/c133606/kexuecheng.shtml

50. www.h2o-china.com (2005) Chengdu: strengthening comprehensive management of urban water environment. 成都:加强城市水环境综合治理>. Available from: https://www.h2o-china.com/news/33601.html
51. www.xy.gov.cn (2021) Learn from the experience of Chengdu to build a harmonious and unified park city of "people, city, environment and industry". 学习借鉴成都经验,建设“人、城、境、业”和谐统一的公园城市>. Available from: http://www.xy.gov.cn/001/001001/20210527/0e2c212b-a09c-4e6c-9696-4b9cf30b3fc9.html
52. Wu D, Zou C, Cao W, Liu L (2018) Analysis of the ecosystem soil conservation function based on the major function-oriented zones across the Yangtze river economic belt, China. Sustainability 10(10):3425
53. Xie L, Cheshmehzangi A, Shao M, Zhang Y, Chan F (2023) Urban sustainability experiments in China: plural approaches for transformation. Handbook on China's urban environmental governance. Edward Elgar Publishing, Cheltenham, pp 169–185
54. Xie L, Flynn A, Tan-Mullins M, Cheshmehzangi A (2019) Water and land: Environmental governance and Chinese eco-development. J Clean Prod 221:839–853
55. Xu B (2023) Transforming China's national new areas towards ecological civilization: a planetary ecological perspective and a scenario modeling method. Doctoral dissertation, University of Pennsylvania
56. Ye SH (2020) Research on Chengdu cultural and creative Industrial Park. 成都文化创意产业园区研究>. Available from: https://max.book118.com/html/2020/0121/8136132134002073.shtm
57. Zeng DZ (2013) China's special economic zones and industrial clusters: success and challenges. Lincoln institute of land policy. Available from: https://www.lincolninst.edu/app/uploads/legacy-files/pubfiles/2261_1600_Zeng_WP13DZ1.pdf
58. Zeuthen JW (2018) Whose urban development? changing credibilities, forms and functions of urbanization in Chengdu, China. Land Use Policy 79:942–951
59. Zhang M, Wen T (2022) The rise of Chengdu between geopolitics and geo-economics: city-regional development under the belt and road initiative and beyond. Trans Inst Br Geogr 47(4):971–989
60. Zhong HL (2015) Chengdu will be built into a major comprehensive transportation hub in western China. 成都构建西部综合交通主枢纽>. Available from: http://paper.ce.cn/jjrb/html/2015-11/02/content_280909.htm
61. Zou J, Sirisuk M (2023) Shu brocade at Sichuan: reinvention of tradition and commoditization in the context of Chinese modern society. Doctoral dissertation, Mahasarakham University

Chapter 8
Nanjing: A Confluence of History, Innovation, and Modernity

1 General Information and Introduction to Nanjing

Nanjing, a city with a rich history spanning over two millennia, occupies a unique and revered place in China's past and present. As the capital during several dynasties, including the Ming and early Qing, Nanjing's historical significance is deeply embedded in its cultural and architectural heritage [14]. The imposing Ming city walls, still standing today, serve as a powerful symbol of its imperial legacy and historical majestic appearance and position. Nanjing's transformation from an ancient capital to a vibrant modern metropolis is marked by its strategic location in eastern China along the Yangtze River [18]. This advantageous position has established Nanjing as a vital economic and cultural hub, integral to fostering trade and commerce in the region [20]. The city's role in facilitating cultural exchanges and economic activities underscores its continuing relevance and dynamism.

The historical narrative of Nanjing is also marked by profound tragedy, particularly the events of the Nanjing Massacre during World War II. This solemn chapter in the city's recent history has left an enduring mark on the collective memory of its residents and the nation. Thus, there is a national memory attached to the city's hardest times in the twentieth century [17]. The Memorial Hall for the Victims of the Nanjing Massacre stands as an emotional reminder of the city's resilience and its unwavering commitment to remembering and honouring those who suffered.

In contemporary times, Nanjing has emerged as a beacon of education and innovation. The city is renowned for its prestigious academic institutions and cutting-edge research centers, which have positioned it at the forefront of intellectual and technological progress. The harmonious blend of ancient monuments and modern advancements in Nanjing exemplifies its ability to integrate its rich heritage with forward-looking aspirations. Economically, Nanjing possesses thriving industries and robust infrastructure, contributing significantly to its status as an economic powerhouse. The city's economic strength is particularly evident in key sectors such as electronics, automotive manufacturing, and petrochemicals. Additionally, Nanjing's innovative

A. Cheshmehzangi and T. Tang, *30 Years of Urban Change in China's 10 Core Cities*, Urban Sustainability, https://doi.org/10.1007/978-981-97-8846-0_8

Fig. 1 Nanjing's central areas and old town in 2005 (map is extracted from open and accessible maps on Google Earth, generated in August 2024)

spirit is highlighted by events such as the annual Nanjing Tech Week, which showcases the city's advancements in technology and entrepreneurship and attracts global attention.

Moreover, Nanjing's urban landscape is directly attributed to its continuous growth and development [15]. Modern skyscrapers and bustling commercial districts coexist with serene parks and historical sites (Figs. 1 and 2), offering a unique blend of old and new. The city's extensive transportation network, including high-speed rail and an international airport, further cements its role as a major transportation hub in the region (see Figs. 3, 4, 5, 6 and 7). Nanjing's cultural richness is also reflected in its numerous museums, theatres, and festivals, which celebrate its diverse heritage and contemporary achievements. The city's culinary scene, known for its traditional Jiangsu cuisine, adds to its cultural appeal, drawing food enthusiasts from around the world. In summary, Nanjing's historical depth, economic vitality, and cultural vibrancy make it a city of great significance in China [5]. Its ability to honour its past while embracing the future positions Nanjing as a unique and enduring metropolis, pivotal to both national history and modern development.

1.1 *Nanjing's Physical and Main Development Over the Past 30 years*

With the total economic output surpassing nine consecutive hundred billion steps, Nanjing's GDP has experienced a remarkable ten-year surge, rising from 730.654 billion yuan in 2012 to 1.63532 trillion yuan in 2021. This represents a significant economic milestone. After adjusting for similar prices, the GDP of the entire city of Nanjing increased by an astounding 104.2% cumulatively between 2013 and 2021, averaging 8.3% annual growth rate, surpassing the average growth rate of the

Fig. 2 Nanjing's central areas and old town in 2024 (map is extracted from open and accessible maps on Google Earth, generated in August 2024)

Fig. 3 Nanjing development in 1984 (map is extracted from open and accessible maps on Google Earth, generated in April 2024)

province by 0.9% points. Nanjing's GDP rose one spot from the three years prior to rank thirteenth nationally in 2012. Nanjing jumped two spots to come in at number eleven by 2014, a position it held for seven years. With a GDP of more than one trillion yuan in 2016, Nanjing cemented its status as one of China's economic superpowers [6].

Nanjing's economic output finally made it into the top ten nationally for the first time since the reform and opening up era in 2020, capping years in which it continuously placed among the top five cities in terms of growth rate. Upon closer inspection, Shenzhen and Nanjing have demonstrated exceptional performance in the last ten years when it comes to the cities that have risen in the national economic rankings. Major cities like Tianjin, Suzhou, and Guangzhou were eclipsed by Shenzhen, and Wuxi, Qingdao, and Tianjin were eclipsed by Nanjing. Nanjing's remarkable GDP

Fig. 4 Nanjing development in 1994 (map is extracted from open and accessible maps on Google Earth, generated in April 2024)

Fig. 5 Nanjing development in 2004 (map is extracted from open and accessible maps on Google Earth, generated in April 2024)

growth per capita serves as more evidence of its rise. With an average annual growth rate of 7.1%, Nanjing's per capita GDP nearly doubled from RMB 85,695 to RMB 174,520 between 2012 and 2021, outpacing the provincial average by 0.3% points. With a per capita GDP of $20,000 in 2017, it surpassed the threshold of a moderately developed economy, rising from $13,575 to $27,051 in USD terms. Nanjing's GDP per capita jumped to fourth place nationally by 2021 [6].

The stability of Nanjing's industrial base has played a significant role in the city's impressive economic rise. In the past, Nanjing has been a highly industrialized city, producing more than half of its industrial output in important industries like electronics, petrochemicals, iron and steel, and automobiles. This growth was made possible by the heavy industries' explosive development in the mid-1990s and early 2000s. Since 2011, Nanjing's economy has shifted to become more dependent on

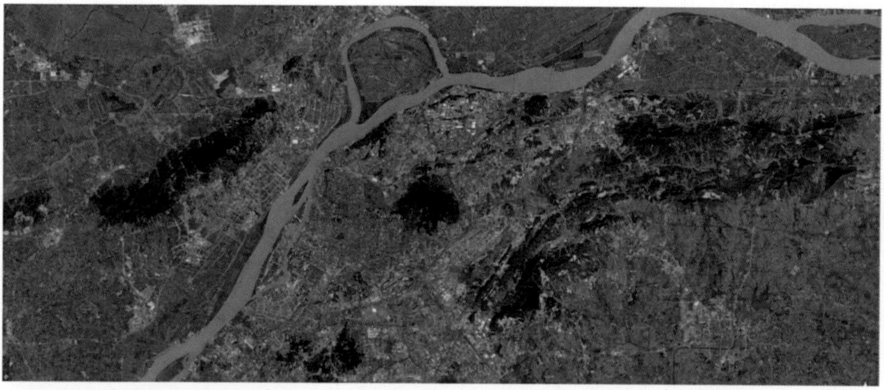

Fig. 6 Nanjing development in 2014 (map is extracted from open and accessible maps on Google Earth, generated in April 2024)

Fig. 7 Nanjing development in 2024 (map is extracted from open and accessible maps on Google Earth, generated in April 2024)

technology, with industries like fiber optics, smart grids, and industrial robotics seeing rapid expansion. The city's industrial value-added growth over the last ten years has averaged almost 8% per year, rising by 10% in 2021 alone [6].

The modern service sector and software industries have also been vital to Nanjing's economic growth. With the software sector at the forefront, the tertiary sector's GDP contribution exceeded 60% by 2017. Nanjing's Software Valley, which employs more than 300,000 people, has developed into a center of innovation. The software and information services sector experienced a sharp increase in revenue, rising from 194.4 billion yuan in 2012 to 720 billion yuan in 2021. Since then, Nanjing has developed sizable industrial clusters; its smart grid and software sectors have been acknowledged as national hubs for advanced manufacturing [6].

Nanjing has demonstrated a clear shift towards technology and innovation through its dedication to research and development. The R&D spending of the city increased at an average annual growth rate of 12.1%, from 20.997 billion yuan in 2012 to 58.879 billion yuan in 2021. R&D spending increased from 2.87% of GDP in 2012 to approximately 3.6% by 2021. Ranking fourth on the Ministry of Science and Technology's list of national innovative cities and 18th in the Global Innovation Index, Nanjing's innovation drive was further acknowledged in 2021 when it was formally approved as a leading national innovative city. The city has effectively shifted into a technology-driven economy, as evidenced by the 7800 high-tech businesses and the high-tech industries' output value, which currently makes up over half of the city's industrial output [6].

2 Nanjing's Development Strategies, Main Policies, and Directions

Nanjing's future urban planning is strategically focused on sustainable growth, aiming to control the resident population at 13 million by 2035. The city's development plan defines Nanjing as "the capital of Jiangsu Province, a national central city, a city rich in national history and culture, an international comprehensive transportation hub, a global innovation highland, and a model city of socialist modernization that satisfies its people." This vision is supported by a comprehensive 35-item indicator system across three key dimensions: bottom-line control, spatial structure efficiency, and quality of life. By 2035, the plan targets an urbanization rate of 92% and a per capita GDP comparable to that of developed cities worldwide. Significant progress toward building a model socialist modernization city is expected by 2025, with the full realization of this goal by 2035.

The urban structure will include the establishment of four municipal centers and the integration of the new Jiangbei main city into the central urban area. The overall spatial layout will follow principles of ecological safety, efficiency, and distinctiveness, forming a balanced pattern of "north–south idyll, central city, river development, and urban–rural integration." This will result in a town system comprising "1 central city, 3 sub-cities, 9 new cities, 29 new towns, and several new communities." The central city will consist of the main city of Jiangnan and the new main city of Jiangbei, while the sub-cities of Liuhe, Lishui, and Gaochun will focus on ecological orientation, industry-city integration, and enhancing regional functionality. The new towns, such as Lukou and Zhetang, will serve as industrial hubs with a focus on strategic emerging industries and balanced job-residential development.

The plan also emphasizes the development of a comprehensive transportation network, aiming to establish Nanjing as an international gateway and a benchmark for green transportation. The city will enhance its role as a national railroad hub, creating a "meter-shaped" high-speed railway network centered on Nanjing, connecting eight directions with twelve lines. This will be complemented by a dual-airport hub pattern, including Lukou International Airport and Ma'an Airport. Additionally, Nanjing will develop a national-level coastal hub port for river-sea intermodal transportation, with plans for four comprehensive port areas along the Yangtze River, capable of handling a container throughput capacity of approximately 9.6 million TEU by 2035.

Urban transit will be further developed with a "three-axis, ten-corridor" urban rail express line network and a "nine vertical and eight horizontal" urban rail transit network, totalling about 1260 km. This will support urban spatial agglomeration and the intensive use of land, facilitating efficient movement within the Nanjing Metropolitan Area. Furthermore, the rational layout of river crossings will be ensured with the planning and pre-control of 25 to 28 crossings, including urban roads and railways, to enhance the city's comprehensive urban integration strategy [12].

3 Example of Nanjing Hexi Central Business District (CBD)

Located in the central region of Nanjing Hexi New City, the CBD spans approximately 4 km from north to south and occupies an area of 2 square meters. Planning and construction for the project began in 2002, with a total floor area of approximately 6 million square meters and a total investment of approximately 30 billion yuan, split into two phases. Situated in the heart of Nanjing Hexi New City, the CBD is 2 square meters in size and extends about 4 km from north to south. The project, which has a total floor area of about 6 million square meters and a total investment of about 30 billion yuan divided into two phases, started planning and construction in 2002. Nanjing Hexi CBD is a collection of finance, headquarters, exhibition, culture and sports, trade, and other modern service industry clusters in Jiangsu Province. It is based in Nanjing, echoing Shanghai and rapidly integrating into the Yangtze River Delta economic circle (see Figs. 8, 9 and 10).

As of 2010, the first phase of the CBD had been finished completely. It covered an area of one square kilometer and had a total length of about two kilometers. It included sixteen iconic buildings and thirteen projects totalling 2.41 million square meters of floor space, with the primary functional modes being business offices, hotels, serviced apartments, and commerce. The Central Park, Administration Square, and Europe City Business Square are three landscape green squares and oasis landscape belts connected by the 2000-m-long central green axis, which perfectly reflects the integration of contemporary business and natural ecology. With a focus on the provision of high-quality goods, hotels, catering, and entertainment services, nearly 500,000 square meters of commerce and trade, primarily in Yanheng, Lianqiang, Zhongtai, Jiaye, and Central Park, as well as other carriers, are opening one after another. This area will eventually become a high-end fashion consumption concentration area in Nanjing and even the metropolitan area.

2010 will see the complete implementation of the CBD Phase I project, which will renovate and upgrade the project as well as further improve the overall quality and

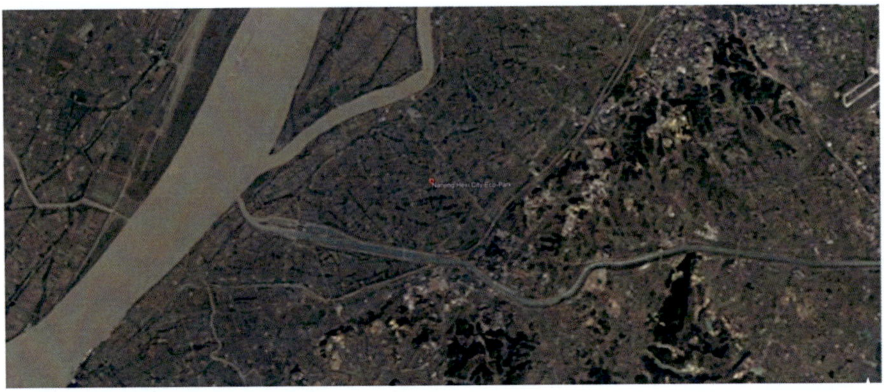

Fig. 8 Nanjing's Hexi New City, the CBD—in 1985 (map is extracted from open and accessible maps on Google Earth, generated in August 2024)

Fig. 9 Nanjing's Hexi New City, the CBD—in 2006 (map is extracted from open and accessible maps on Google Earth, generated in August 2024)

functionality of the CBD. Preliminary statistics show that Hexi CBD has attracted over 1400 businesses of all kinds, including over 100 financial enterprises, over 60 headquarters and regional headquarters of businesses [16], over 60 high-tech businesses, and over 10,000 workers in the regional office, including over 50 individuals with high-level innovation and entrepreneurial talent. The concentration of these elite businesses and elite individuals has emerged as a major catalyst for the CBD's explosive growth.

The second phase of the CBD's construction in the west of the river is moving forward quickly in tandem with the Youth Olympic Games in Nanjing. Other projects that have started construction include the Jiangsu Provincial Banking Regulatory Bureau, the Securities and Futures Bureau office building, Hong Kong Sun Hung Kai, the Nanjing World Trade Center, the 14 Guo Rui Building, the provincial Bank

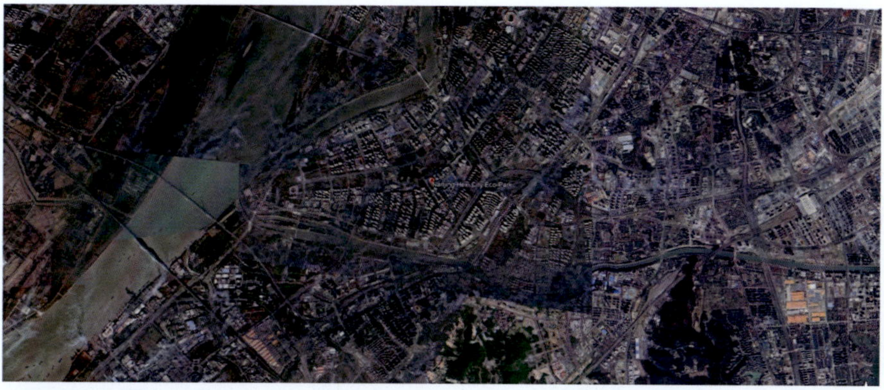

Fig. 10 Nanjing's Hexi New City, the CBD—in 2024 (map is extracted from open and accessible maps on Google Earth, generated in August 2024)

of Communications, Xinhua Newspaper, Mobile Nanjing Branch, Unicom Jiangsu Branch, Huatai Securities, Nanjing Newspapers, Guotai Junan, Jiangsu Credit Union, and the comprehensive elements of the market west of the river. The project, which has a total investment of over 20 billion yuan and a construction area of 3.5 million square meters, aims to create a financial insurance and headquarters agglomeration area. Prosperity will drive the project's popularity, and it is anticipated that the CBD Phase II will be finished and delivered by 2014.

The Party Committee and government of Jianye District, Nanjing proposed the "One District, Five Cities" industrial plan, which focuses on the development of the modern service industry, including modern finance, information service, business office, tourism and exhibition, culture and sports, and leisure, as well as related ancillary services, with the goal of creating an area of regional significance [10]. The Hexi CBD is responsible for carrying out the overall construction and implementation tasks of the financial city, exhibition city, and culture and sports city in the modern service industry cluster. In order to support the economic growth of the Nanjing urban area and propel the development of the Hexi New City industrial system, it focuses on the development of modern finance, information services, business offices, tourism and exhibition, culture, sports, and leisure, and related supporting services. It also aims to create service industry clusters with regional characteristics, forming the core area of the new city center. With five to ten years of work, Hexi CBD, the window area of the 2014 Youth Olympic Games main venue, hopes to establish itself as a unique, nationally renowned, and first-class financial and headquarters economic center located in Shanghai.

According to Baidu Encyclopedia (n.d.), Hexi Central Business District (CBD) in Nanjing is becoming a hub for a variety of businesses these days, including Fortune 500 companies, listed companies, and multinational corporations as well as regional headquarters. Modern infrastructure, state-of-the-art business amenities, and innovative urban planning concepts are driving this influx, establishing Hexi CBD as a center of contemporary commerce. The district's strategic location contributes to its allure by facilitating seamless connectivity and accessibility—two factors that are critical for businesses operating on a global scale. Because of this, Hexi CBD is quickly becoming a representation of Nanjing's modernization and internationalization, where a lively, multicultural urban atmosphere is developing. Other examples of mixed-use development have followed a similar pattern of physical urban development in Nanjing (see Figs. 11, 12 and 13 as an example).

Fig. 11 Nanjing's new mixed-use development in 2006 (maps are extracted from open and accessible maps on Google Earth, generated in August 2024)

Fig. 12 Nanjing's new mixed-use development in 2012 (maps are extracted from open and accessible maps on Google Earth, generated in August 2024)

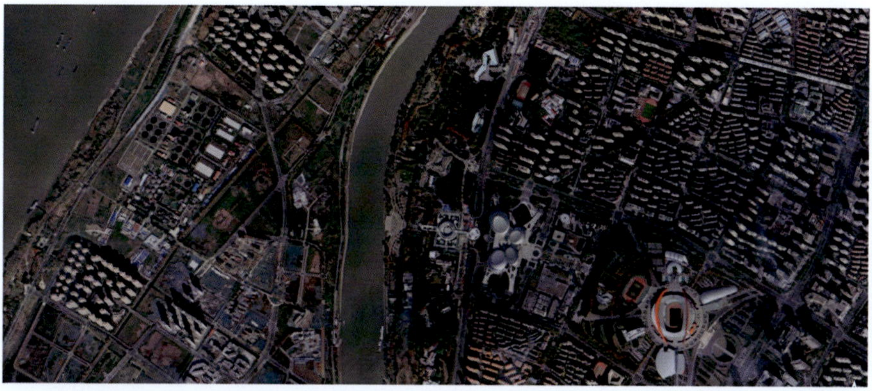

Fig. 13 Nanjing's new mixed-use development in 2024 (maps are extracted from open and accessible maps on Google Earth, generated in August 2024)

4 The Local Government's Projection for the Future Development of Nanjing

The local government is adamant about boosting domestic demand, controlling foreign demand, and concentrating on fortifying and stabilizing the economic recovery's upward trend. We will cultivate new dynamics in international trade more quickly, better integrate investment and consumption, and consistently improve the endogenous dynamics and vitality of economic development. This means that in the future, development will stick to the creation of strong, clustered, and long chains as well as work to support the development of significant industrial breakthroughs. aggressively encourage new industrialization, carefully carry out the three-year action plan to bolster the city's industry, enhance the "seven one" manufacturing and "six one" service industries, work to advance the system through special classes and projects, and encourage the development of a contemporary industrial system.

Furthermore, the government wants to follow sensible and strong four-chain integration, emphasizing the development of an industrial science and technology innovation center's primary carrying area with a global impact. We will encourage the seamless flow and deep integration of the innovation chain, industrial chain, capital chain, and talent chain and hasten the formation of new productivity, taking the comprehensive reform of the science and technology system as our model [13]. This means that in order to move forward, development must intensify reform, broaden opening-up, and concentrate on creating a hub that connects the domestic and international double cycles. We will effectively leverage the "key move" of opening-up and reform to mobilize elements and resources on a larger scale and to foster endogenous power at a deeper level.

The plan states that the government will uphold planning leadership, empower people digitally, and work to enhance the standard of urban functions. We'll put the idea of the "people's city" into reality, change the way megacities develop more quickly, and build livable, resilient, and intelligent cities. This calls for continuing to uphold passing down traditions and fostering new ones while concentrating on creating a powerful socialist cultural metropolis. We will brighten the golden signs indicating the capitals of literature, green humanity, and fraternity, and we will quicken the building of a quaint Nanjing that is more well-known throughout the world and has a thicker culture [13].

Adhering to ecological priority and green development, as well as working to encourage the peaceful coexistence of man and nature, are additional important strategies [1]. The local government will put President Xi Jinping's concept of an ecological civilization into practice, work together to encourage pollution and carbon reduction, growth and expansion of green spaces, and quicken the building of a stunning Nanjing. As part of its development plan, the government will keep pushing for the integration of urban and rural areas, supporting agriculture and helping farmers, and working to advance the comprehensive revitalization of rural areas. We intend to construct a robust urban modern agricultural city that embodies the essence of Nanjing, expediting the dissemination of a novel vision for the modernization of rural and agricultural areas [13].

Additional areas that warrant further examination are those that align with development patterns that prioritize system optimization, financial stability, and the preservation and enhancement of people's standard of living. concentrating on reaching a greater level of "seven people's livelihoods," acting more to improve people's lives and cheer them up, and successfully resolving the "urgent, difficult, worrying and hopeful" issues that the public faces. A significant change will present an opportunity to uphold methodical reasoning, scientific policymaking, and the modernization of the urban governance system and governance capacity. Enhance the public safety system, address the deficiencies in urban governance, and encourage a dynamic equilibrium that harmonizes development and security [13].

4.1 Challenges of the Past and Continuing Issues for Consideration

Nanjing is confronted with formidable obstacles in its pursuit of preserving its economic growth and tackling new problems. One of the main issues is the low effective demand, which has caused a slower-than-anticipated recovery in consumption when combined with low social expectations. The growth of foreign trade is still under pressure, and cyclical factors disproportionately affect the petrochemical and steel industries. To remain competitive in the global market, important industries like the automotive and electronic information sectors must accelerate their transformation.

The city has made progress, but there is still a lack of support for strategic emerging industries, which has caused some profitable service sectors to slow down in their growth [8, 9]. Companies are having more trouble running their operations, which makes it harder to keep the economy stable. In order to maintain long-term growth, high-level innovation platforms are still insufficient, and research and development in important core technologies must be accelerated. Nanjing is also unable to fully realize its potential for innovation since the translation of scientific and technological advancements into useful applications is still lagging behind.

Sustainability of the environment is another issue. Although there has been improvement, there are still gaps in the availability of high-quality public services, and the ecological environment's quality is still below public expectations [4]. Significant risk factors are also present in some areas, which calls for an immediate resolution to raise the city's overall intrinsic safety level [7]. To ensure that Nanjing can prosper in the face of future uncertainties, addressing these issues will call for more detailed and accurate business services in addition to improved governance [13].

4.2 Key Lessons Learnt from the City for Other Cities or Contexts

Other cities can learn a lot from Nanjing's urban development strategy about how to build resilient, dynamic, and sustainable environments. Emphasizing the value of a well-balanced urban mix that incorporates green areas, historical sites, and contemporary infrastructure is a crucial lesson. This strategy not only makes the city more aesthetically pleasing, but it also promotes a peaceful way of life. The city of Nanjing has placed a strong emphasis on sustainable development, as evidenced by the establishment of ecological parks and ecologically sensitive urban design. This highlights the need to integrate environmental stewardship with growth goals.

A crucial element of Nanjing's approach is the smooth amalgamation of business, residential, and leisure areas. Comprehensive urban planning creates a well-rounded urban experience for both locals and visitors by promoting a dynamic city life where housing, employment, and recreational opportunities coexist. A distinctive sense of place that fortifies ties between neighbours and draws in cultural tourists is provided by the integration of cultural landmarks into the urban landscape, which enhances the city's identity and cultural vibrancy [3]. The extensive Metro system in Nanjing serves as an example of the city's investment in effective transportation infrastructure, highlighting the significance of mobility in promoting economic growth and raising living standards. In addition to making it easier for people and goods to move around, efficient transportation networks also lessen traffic and pollution, which helps create a more sustainable urban environment. Repurposing land strategically can result in vibrant community assets. One example of this is the redevelopment of Xuanwu Lake Park into a recreational hub [19]. This strategy not only improves public areas but also increases the city's recreational value, enhancing the well-being of its citizens and drawing tourists.

Finally, Nanjing's emphasis on promoting economic diversification offers resilience as a model. The city can more effectively withstand economic shocks and maintain long-term growth by fostering a diverse economic base. In order to create thriving cities, these Nanjing lessons highlight the necessity of an all-encompassing approach to urban development that strikes a balance between infrastructure, environmental sustainability, cultural enrichment, and economic diversification.

References

1. Allam Z, Cheshmehzangi A, Jones DS (2023) Crafting and transforming colonial empires and landscapes. In: Climate and social justice: the political economy of urban resilience and mercantilism. Singapore, Springer Nature Singapore, pp 61–75
2. Baidu Encyclopedia (n.d.) 南京河西中央商务区. Available from: https://baike.baidu.com/item南京河西中央商务区/5336103
3. Cheshmehzangi A (2023) Urban regeneration and creating urban landscapes: enhancing city life experiences in Lijiang. In: Mapping urban regeneration: city life experiences in Yunnan, China. Singapore, Springer Nature Singapore, pp 139–169
4. Cheshmehzangi A, Dawodu A, Sharifi A (2021) Sustainable urbanism in China. Routledge, New York
5. Cheshmehzangi A, Tang T (2022) Yangtze river delta city cluster: from integrated production and services to digital technology innovation. In: China's City Cluster Development in the Race to Carbon Neutrality. Singapore, Springer Nature Singapore, pp 47–62
6. China Jiangsu Net (2022) Nanjing: A trillion-dollar city's road to catch up. <<南京:一座万亿城市的追赶之路>>. Available from: https://jsnews.jschina.com.cn/nj/a/202206/t20220627_3025405.shtml
7. Fu Y, Yang X, Wang T, Supriyadi A, Cirella GT (2022) Spatial pattern characteristics of the financial service industry: evidence from Nanjing, China. Appl Spat Anal Policy 15(3):595–620
8. Gao J, Chen W (2020) Spatial restructuring and the logic of industrial land redevelopment in urban China: III. a case study of the redevelopment of a central state-owned enterprise in Nanjing. Cities 96:102460
9. Gao J, Yuan F (2017) Economic transition, firm dynamics, and restructuring of manufacturing spaces in urban China: empirical evidence from Nanjing. Prof Geogr 69(3):504–519
10. JLL (2022) Industrial growth burgeoning in Nanjing. Available from: https://research.jllapsites.com/industrial-growth-burgeoning-in-nanjing/
11. Lipkin Z (2006) A new society for a new capital. In: Useless to the state. Harvard University Asia Center, pp 21–60. Available at: https://brill.com/downloadpdf/book/9781684174263/BP000003.pdf
12. Nanjing Government (2022) Nanjing Land and Space Master Plan (2021–2035) <<南京市国土空间总体规划 (2021–2035年)>> Available from: https://www.nanjing.gov.cn/njxx/202210/t20221029_3740005.html
13. Nanjing Government (2024) 2024 Nanjing municipal government work report. << 2024年南京市政府工作报告 >> Available from: https://www.nanjing.gov.cn/zdgk/202403/t20240311_4183006.html
14. Rimmer P, Dick H (2013) The historical dimension. global city challenges: debating a concept, improving the practice. Palgrave Macmillan UK, London, pp 63–87
15. Shen J, Kee G, Shen J, Kee G (2017) Nanjing: city positioning and development strategies in the twenty-first century. In: Development and planning in seven major coastal cities in southern and eastern China, pp 181–209
16. Song S (2022) Multiple meanings of everyday spaces: understanding (liveable) streets in contemporary China. Doctoral dissertation, University of Sheffield
17. Wang L (2016) Becoming Urban in the Chinese Way: The Politics of Planning and Urban Change in Nanjing, China. Doctoral dissertation, The Ohio State University

18. Yao Y, Yao Y (2016) Nanjing historical landscape and city changes. Nanjing: historical landscape and its planning from geographical perspective, pp 39–114
19. Yuan T (2011) Landscape planning of urban lake park in China a case study on Nanjing. Available from: https://www.diva-portal.org/smash/record.jsf?pid=diva2:829183
20. Zhaoyan Y, East DC (2021) People of Nanjing: a cultural perspective on a historic Chinese city. Available: https://www.torrossa.com/it/resources/an/5189514

Chapter 9
Wuhan: A Blueprint for Urban Innovation, Resilience, and Sustainable Development

1 Introduction to Wuhan City, China

Wuhan, a city steeped in history, is a symbol of China's lasting legacy, capturing the spirit of resiliency and rejuvenation that has characterized the country's progress over centuries [10]. Wuhan, a city in central China, is significant for reasons that go well beyond its location [18]. It has long been a hub for cross-cultural interchange [3], where the influences of many dynasties met to weave a rich tapestry of customs and inventions that have shaped the city and the surrounding area.

Wuhan was historically a hub for trade and transportation due to its advantageous location at the meeting of the Yangtze and Han rivers [9]. Wuhan was able to grow as a center of economic activity thanks to its advantageous location, which made it easier for people and goods to travel across China [18]. From the ancient markets of antiquity to the contemporary industrial complexes that now dot its landscape, the city has always been an essential component of the country's economic network.

Wuhan's historical story is intricately linked to its significance at pivotal junctures in Chinese history. This is where the 1911 Wuchang Rebellion lit the spark for revolution, which ultimately resulted in the fall of the Qing Dynasty and the founding of the Republic of China. Wuhan gained national prominence as a result of this historic event, which changed the course of history and ushered in a new era for China.

Wuhan is a major economic force in the modern era, and its significance cannot be emphasized. The city has developed into a significant hub for industry, trade, and technological advancement [18]. A wide range of industries, such as biotechnology, steel manufacturing, and auto manufacturing, define its economic landscape. Wuhan has a significant global economic impact due to the products and technologies it produces that are sold in markets all over the world [8]. Due to its advantageous location, the city is also a major participant in China's Belt and Road Initiative, serving as a hub for foreign investment and trade.

A. Cheshmehzangi and T. Tang, *30 Years of Urban Change in China's 10 Core Cities*, Urban Sustainability, https://doi.org/10.1007/978-981-97-8846-0_9

Wuhan is well known for its intellectual contributions and vibrant culture in addition to its economic strength. Famous sites like East Lake, one of China's largest urban lakes and a tranquil haven from the busy city, and the Yellow Crane Tower, a testament to Wuhan's lasting legacy, serve as examples of the city's rich cultural legacy. These cultural assets not only bear witness to Wuhan's historical significance but also act as inspiration for both locals and tourists. Wuhan's reputation as a significant national and international city is further strengthened by its position as a major hub for research and education. Wuhan University is one of the most prominent universities in China, with a beautiful campus that overlooks East Lake [18]. The university is essential in forming Wuhan's intellectual landscape, as are other academic and research institutions in the city. The city attracts academics, researchers, and students, creating a thriving academic community that advances knowledge and creativity.

Furthermore, Wuhan's adaptability and resiliency have been tested recently, especially during the COVID-19 pandemic. The city's handling of the outbreak has demonstrated not only its ability to handle crises but also its status as a global city with connections to the rest of the world that affect both local and global affairs [18]. The people of Wuhan are resilient and dedicated to rebuilding and progressing, as evidenced by their recovery and subsequent renaissance as a vibrant metropolis.

Wuhan is essentially a city where the past and present collide to produce a singular fusion of modernity, culture, and history. Its history as a hub for intellectual inquiry, a cradle of cross-cultural exchange, and a lighthouse of economic advancement continues to define it and shape its future. Wuhan continues to stand as a testament to the inventiveness and tenacity that characterize China's spirit and as a symbol of the country's lasting legacy (see Figs. 1, 2, 3, 4, and 5).

Fig. 1 Wuhan's Development in 1984 (map is extracted from open and accessible maps on Google Earth, generated in August 2024)

Fig. 2 Wuhan's Development in 1994 (map is extracted from open and accessible maps on Google Earth, generated in August 2024)

Fig. 3 Wuhan's Development in 2004 (map is extracted from open and accessible maps on Google Earth, generated in August 2024)

2 Wuhan's 30 Years of Urban Development

Wuhan has greatly accelerated its growth as a national center city over the last ten years, achieving previously unheard-of levels of comprehensive strength. The city has seen incredible economic growth, setting new standards in a number of industries and establishing itself as a major player in China. With an average annual growth rate of 7.4%, Wuhan's regional GDP reached an impressive 1.77 trillion yuan by 2021, increasing its economic output by 2.3 times over that of 2012. Thanks to this outstanding accomplishment [18], Wuhan is now the most economically productive city in central China and among the top ten cities in China overall. Remarkably, ten of its districts now have economies producing more than 100 billion yuan apiece. Additionally, Wuhan's per capita GDP has increased significantly, from 12.6 million yuan

Fig. 4 Wuhan's Development in 2014 (map is extracted from open and accessible maps on Google Earth, generated in August 2024)

Fig. 5 Wuhan's Development in 2024 (map is extracted from open and accessible maps on Google Earth, generated in August 2024)

to over 20 million US dollars, demonstrating the city's thriving economy. Wuhan's increasing economic appeal has drawn a lot of development and investment, which has helped the city grow into a bustling metropolis. Over 7 trillion yuan have been invested in fixed assets in the city, growing at an average annual rate of 8.1%. There are currently more than 1.7 million market entities in the city, which has carried out 850 significant municipal initiatives. Additionally, 309 of the top 500 global corporations have invested in and established a presence in Wuhan [18], demonstrating the city's attraction to multinational corporations. In addition, there has been a significant rise in the resident population, which increased from 10.14 million in 2012 to 13.65 million in 2021 (a net increase of 3.59 million, or a cumulative growth rate of 34.6%).

Wuhan has aggressively supported the creation of the Wuhan metropolitan area over the last ten years, establishing a new paradigm for coordinated regional development. With the completion of the third phase of Metro Line 11, which includes the Ezhou Gedian section, and the operation of five intercity railroads, the city has made great progress in enhancing its transportation infrastructure. Additionally, 50 new roads have been built as a result of efforts to relieve infrastructure bottlenecks, and 410 services are now offered across municipal boundaries thanks to improvements made to cross-city government services [18]. The city's commitment to high-quality urban development is demonstrated by the increase in the resident population's urbanization rate, which rose from 78.98% in 2012 to 84.56% in 2021 as a result of these initiatives.

Wuhan's ability to develop has reached unprecedented levels thanks to its commitment to creating a cutting-edge industrial system [11]. With the help of advanced manufacturing, modern service industries, and strategically important emerging industries, the city has successfully established a "965" modern industrial system. Wuhan has concentrated on developing important industries that have the potential to generate trillions of yuan, including biotechnology and health, next-generation information technology, and automotive manufacturing and services. With the service sector's value-added output surpassing one trillion yuan and its share of GDP rising by 11.1 percentage points to 62.5%, the industrial structure has undergone further upgrading. Six service industry clusters have been designated as national demonstration zones, and Wuhan currently has 15 businesses listed among China's top 500 service industry companies [7]. High-tax buildings have also become more prevalent in the city,83 of these buildings generate more than 100 million yuan in tax income annually [18].

All of these accomplishments show how far Wuhan has come in the last ten years. In addition to strengthening its industrial and economic capacities, the city has also improved its standing as a major metropolitan area, laying the groundwork for future expansion and advancement.

2.1 Wuhan and Its Development Strategies

By 2035, Wuhan will have achieved its goal of becoming a national center city, firmly establishing itself as the center of the Yangtze River Economic Belt and greatly raising its profile as a global city [17]. By then, Wuhan will have made significant strides toward modernization, demonstrating gains in its overall competitiveness, scientific and technological prowess, and economic strength. It is anticipated that the city's overall economic output will soar to new heights and that both urban and rural citizens' per capita incomes will have increased significantly. Wuhan is a forward-thinking city that strives to lead the country in technological innovation and creative industry standards.

Wuhan will be clearly transformed into a global market hub by 2035, with a significant rise in its degree of openness to the outside world [17]. By then, the city will have effectively integrated information technology, urbanization, new industrialization, and agricultural modernization to create a comprehensive and advanced economic system [4, 22]. Together with this economic development, the city's capacity for governance and governance system will be significantly improved [21]. Wuhan is dedicated to making sure that the rights of its citizens to equal participation and development are completely upheld by establishing a society, government, and legal system based on the rule of law and embodying justice and fairness.

Wuhan is planned to become a hub for talent, education, culture, health, and education in tandem with these economic and governance developments. In order to maintain the highest standards of public safety and security, the city intends to build a Safe Wuhan by going above and beyond [17]. Focusing strongly on cultural and societal development will improve residents' quality of life, and Wuhan's urban culture's soft power will be greatly increased. Wuhan will be positioned as a leading city in terms of cultural influence and intellectual capital due to the increased level of civility and improved quality of civilization among its citizens.

Wuhan's dedication to sustainability will also be highlighted, as eco-friendly production and lifestyle choices become the standard. The city hopes to significantly enhance its natural surroundings by 2035 and become a model city for the "Beautiful China" program. There will be extensive efforts to improve environmental quality, which will help to build a more sustainable and livable urban environment. Policies that support environmentally friendly practices and green technologies will help to support this shift towards sustainability in all areas of the economy.

As Wuhan works to create a more equitable society, the growth of the middle class will be a major priority. All citizens will have equal access to basic public services, including social services, healthcare, and education [17]. New heights in the integration of urban and rural areas will be reached, promoting balanced development and lessening differences between various parts of the city. This strategy will have a noticeable improvement in people's general well-being and quality of life while also promoting the full development of each individual and the prosperity of all citizens.

Wuhan aspires to be a city of innovation and modernity by 2035, where social justice, environmental sustainability, and cultural diversity coexist with economic dynamism. The goal for Wuhan is not only to achieve economic prosperity but also to develop an all-encompassing urban area that benefits all citizens, becoming an example city that other cities can use as a source of inspiration [17]. All facets of society will need to work together to create this future, but the groundwork has already been done and the city is headed for a prosperous and bright future for everybody.

3 Wuhan's Hi-Tech Zone for Innovation and Technologies

Wuhan East Lake New Technology Development Zone, also referred to as East Lake Hi-tech Zone, is also known as China's Optical Valley, or simply Optical Valley. It was established in 1988 and is the nation's first batch of state-level high-tech zones. It is also the second national demonstration zone of independent innovation,

Fig. 6 Wuhan East Lake New Technology Development Zone in 2006 (map is extracted from open and accessible maps on Google Earth, generated in August 2024)

China (Hubei) Pilot Free Trade Zone Wuhan Area, and it has been approved as the nation's bases for the bioindustry, centralized construction of talent and enterprises, the nation's first batch of dual-creation demonstration bases, and so forth (WEHDZ, n.d.). Following over three decades of growth, East Lake Hi-tech Zone has emerged as a significant hub for the industry's advancement.

East Lake Hi-Tech Zone's overall strength and brand influence have significantly increased following more than 30 years of development. It is now one of China's 10 "world-class high-tech parks" and has risen to the top spot among the country's 169 national high-tech zones in terms of knowledge creation and technological innovation (WEHDZ, n.d.). Surrounded by 42 colleges and universities, including Wuhan University and Huazhong University of Science and Technology, 56 national and provincial research institutes, more than 300,000 professionals and technicians, and more than 800,000 college students, the 518 square kilometer East Lake Hi-Tech Zone is one of China's three most intellectually dense zones (WEHDZ, n.d.) (Figs. 6, 7, and 8).

Huashan Street, Jiufeng Street, Baoxiu Street, Binhu Street, Optics Valley Biocity, Wuhan Future Science and Technology City, Wuhan East Lake Comprehensive Bonded Zone, Optics Valley Optical Electronics and Information Park, Optics Valley Modern Service Industry Park, Optics Valley Intelligent Manufacturing Park, Optics Valley China Science and Technology Park, and Optics Valley Center City (WEHDZ, n.d.) are among the eight streets that fall under its jurisdiction.

Currently, the East Lake Hi-tech Zone is pushing the development of the "221" industrial system by building "two strong drives, two new integrations, seize the future". With an emphasis on the "optical core screen end network," life and health are the two main industry clusters [6]. There will be a thorough promotion of the strong chain, which will extend and supplement the chain and further solidify the industrial foundation (WEHDZ, n.d.). Key areas of the short board that require attention include making up for missing core equipment, basic technology, and key basic materials.

Fig. 7 Wuhan East Lake New Technology Development Zone in 2012 (map is extracted from open and accessible maps on Google Earth, generated in August 2024)

Fig. 8 Wuhan East Lake New Technology Development Zone in 2024 (map is extracted from open and accessible maps on Google Earth, generated in August 2024)

The ability of the industry chain supply chain to be independent and controllable will also be improved. Deep integration of the digital economy and the new consumer will be deeply promoted. Numerous new product and technology application scenarios will be built, and new products and technologies will be actively encouraged to be used. The development of 5G, artificial intelligence, the Internet of Things, and other digitalized industries,the forward-looking layout of several future industries; the formulation of support for the cultivation of future industries such as brain science, quantum science and technology, blockchain, and other future industrial policies; the support for the accelerated development of online education, online healthcare, online entertainment, and other digital consumer services; new forms and new modes; and the enhancement of the energy level of digital consumption. Launch a fresh wave of technological enterprise transformation, and quicken the advancement of superior manufacturing digitalization, networking, and intelligence.

The East Lake Hi-Tech Zone's GDP is projected to increase by 4.7% in 2023. These include the GDP as a whole, investments in fixed assets, and investments made in lieu of the entire amount of funds in the city. East Lake Hi-tech Zone will concentrate on building East Lake Science City during the "14th Five-Year Plan" period (WEHDZ, n.d.), striving to create a "city of science, a city of light, a city of yearning." The Optics Valley will be developed into a model area for the middle reaches of the Yangtze River for putting the concept of "two mountains" into practice, a pioneer area for high-quality development in the central region, and a demonstration area for national innovation-driven development (WEHDZ, n.d.). In 2035, the East Lake Hi-tech Zone is expected to have significantly enhanced comprehensive strength, assume the lead in essentially implementing socialist modernization, and establish the "World Optical Valley" first.

4 Wuhan's Local Government Projection for Future Development

The main ambitious targets of Wuhan's economic and social development are the creation of over 250,000 urban jobs, a controlled increase in consumer prices of approximately 3%, and a regional GDP growth of 6%. These goals show a balanced approach to development, emphasizing the preservation of the natural environment while making sure that the income growth of the populace keeps up with economic growth. Wuhan's commitment to achieving provincial goals for emission reduction and energy conservation further highlights the city's commitment to sustainable growth.

In an effort to become a significant hub of innovation in the country, Wuhan is working to advance advanced scientific and technological self-reliance. The city understands that innovation-driven development is essential to its long-term success. Wuhan is dedicated to coordinating initiatives in education, science and technology, and talent development to consistently produce new growth-promoting factors. Wuhan wants to be at the forefront of cutting-edge industries and keep its competitive edge both domestically and internationally by promoting an atmosphere that prioritizes scientific research and technological advancements.

Wuhan is concentrated on advancing new industrialization and hastening the establishment of a contemporary industrial system supported by advanced manufacturing in parallel. The city's approach is to develop new types of superior productivity by utilizing scientific and technological innovation to propel industrial progress [12]. This entails not just modernizing already-existing industries but also encouraging the growth of new ones that will shape the economy of the future [19]. Wuhan wants to make sure that its economy is sustainable and resilient over the long run by developing a strong and diverse industrial base.

Wuhan is also dedicated to fostering supply and demand synergy, hastening the establishment of a positive feedback loop that benefits investment and consumption. Stabilizing expectations and increasing domestic demand—both of which are viewed as essential for long-term economic growth—are highlighted in the city's strategy [19]. Wuhan aims to establish a self-sustaining economic environment where investment and consumption mutually reinforce one another by fostering endogenous drivers of development, such as productivity and innovation.

Important elements of Wuhan's development strategy are also increasing opening-up and advancing reform. In response to the demands of the modern era, the city is resolved to expedite the construction of a new inland opening-up high ground. Wuhan hopes to better serve and integrate into the national and international development patterns by moving forward with comprehensive reforms and high-level opening-up in a coordinated manner. This entails strengthening ties with neighbouring areas, drawing in foreign capital, and establishing itself as a key component of China's larger economic plan.

Wuhan's efforts to improve the city's planning, building, and governance demonstrate the city's dedication to urban development. Accelerating the development of a smart, resilient, and liveable city that satisfies resident needs and is ready for future challenges is the aim [19]. Wuhan is committed to enhancing urban functions through full-cycle management, making sure that the infrastructure, services, and governance of the city are designed with long-term resilience and sustainability in mind (see Figs. 9, 10, and 11 as an example).

The city is also committed to advancing low-carbon, green development and ecological civilization. Wuhan is committed to ecological priority and green development, and it wants to showcase its special qualities as a city of rivers, lakes, and wetlands [19]. The objective is to keep constructing a waterfront ecological city of

Fig. 9 Wuhan's example of mixed-use development in rapidly-developing areas in 2000 (map is extracted from open and accessible maps on Google Earth, generated in August 2024)

Fig. 10 Wuhan's example of mixed-use development in rapidly-developing areas in 2012 (map is extracted from open and accessible maps on Google Earth, generated in August 2024)

Fig. 11 Wuhan's example of mixed-use development in rapidly-developing areas in 2022 (map is extracted from open and accessible maps on Google Earth, generated in August 2024)

international renown that balances urban life and the environment. This entails incorporating sustainable practices into every facet of urban life in addition to safeguarding and improving the natural resources of the city.

Finally, with the goal of quickening the entire revitalization of its countryside, Wuhan is concentrated on the development of agriculture and rural areas as its top priorities. Based on the prosperous "Ten Million Projects" experience, the city is striving to improve rural building, rural industrial development, and rural governance. Wuhan aims to establish a more integrated and balanced urban–rural development pattern by promoting rural revitalization, which will guarantee that the advantages of economic growth are distributed throughout the entire city.

In conclusion, Wuhan's development objectives show a thorough and progressive approach that strikes a balance between social justice, environmental sustainability, and technological innovation in addition to economic growth [19]. The city intends to maintain its position as a major player in China's ongoing development by pursuing a vision of resilience, inclusivity, and global competitiveness.

4.1 Wuhan's Challenges of the Past and Present

Wuhan has shown incredible adaptability and resilience in the face of economic difficulties, but the groundwork for a long-term recovery is still lacking. Although the city has made progress in recovering from recent setbacks, there are still important sectors facing heavy pressures, so the momentum is not uniform. For example, there have been indications of weakness in foreign trade growth and investment, which reflects the larger issues facing both domestic and international markets. Wuhan's industrial base was formerly anchored by the steel and automotive industries, but these sectors are still under a lot of stress. The problems are made worse by overcapacity and changing consumer demands.

Particularly at risk are small and medium-sized businesses (SMEs), which are vital to Wuhan's economy. These companies are vulnerable to protracted periods of instability because they frequently lack the cash reserves and market access necessary to withstand changes in the economy. Thus, it is crucial to find a faster way to transform kinetic energy into sustainable growth. Innovation and industrial upgrading must be given top priority [5] to revive these industries and bring new vitality to Wuhan's economy.

Another crucial area in which Wuhan needs to make major progress is urban renewal. The city has grown so quickly in the last few decades that better infrastructure development and urban management are urgently needed. Wuhan needs to step up its urban renewal efforts to stay competitive and liveable. It should improve public services, modernize its infrastructure, and develop more resilient and sustainable urban areas. Fine management techniques must be implemented to guarantee that these developments meet the highest standards of effectiveness and caliber.

Wuhan also regards environmental and ecological governance as a top priority. Because of its location at the Yangtze and Han river confluence, the city is especially susceptible to environmental problems like pollution, flooding, and habitat degradation. Wuhan has made strides in resolving these problems, but more work has to be done to establish sustainable practices in all areas and strengthen ecological governance. This entails boosting clean energy, developing green infrastructure, and improving the city's natural ecosystems. The intention is to make Wuhan an example of an urban sustainable community, where environmental preservation and economic growth coexist.

Another urgent challenge facing Wuhan is preventing and resolving risks in important areas. The city is beset with covert challenges in a number of areas, such as social inequality, public health hazards, and unstable finances. To keep these risks from jeopardizing the advancement of the city, they must be carefully controlled. Wuhan must make sure that all stakeholders are involved in the city's efforts to become more resilient, as well as strengthen its risk management frameworks and institutional capacity to respond to crises.

Additionally, a gap needs to be closed between Wuhan's citizens' expectations and means of subsistence and the government's offerings. Even though the city has spent a lot of money enhancing public services, there is still a gap between what residents expect and what they get [19]. A portion of this disparity can be attributed to government employees who have not yet completely adjusted to the new developmental era. Wuhan needs to concentrate on creating a more receptive and functional government that can fulfill the demands of its people in order to address this. This entails making the government more open, accountable, and community-engaged in addition to enhancing the abilities and dispositions of public employees.

Wuhan's future course will necessitate a multimodal strategy that takes into account these difficulties as well as the city's advantages. The city needs to keep up its efforts to modernize its infrastructure, modernize its economy, and improve environmental sustainability. Wuhan must simultaneously exercise caution when it comes to risk management, filling in service delivery gaps, and promoting a more inclusive and participatory style of government [19]. By doing this, Wuhan will be able to build on its successes and create a strong basis for continued development and prosperity.

5 Key Lessons Learnt from Wuhan for Other Cities

Wuhan's urban development journey offers insightful lessons for other cities trying to balance growth, innovation, and sustainability. Wuhan's experience highlights the significance of creating zones specifically designated for technological innovation. These areas act as stimulants to draw in talent, encourage entrepreneurship, and boost the economy. Wuhan has been able to establish itself as a leader in high-tech industries and attract talent from both domestic and foreign sources by fostering an environment that encourages innovation. Cities all over the world that want to improve their economic competitiveness in a global market that is becoming more and more reliant on technology can use this innovation hub model.

Wuhan did not, however, achieve technological development success on its own. The city has placed great emphasis on the assimilation of technological innovations with thoughtfully designed urban facilities in an effort to improve the general standard of living for its citizens. This strategy emphasizes how important it is for cities to strike a balance between developing liveable spaces and economic growth. Wuhan has established an environment that facilitates innovation and quality of life by providing green spaces, cultural facilities, and essential services in addition to technological zones. By realizing that sustainable urban development necessitates not only economic vitality but also a focus on resident well-being, other cities can draw lessons from this.

Wuhan also teaches us the importance of establishing collaborations between businesses and academic institutions [20]. The city's technological progress has been greatly aided by these partnerships, especially in fields like biotechnology, information technology, and advanced manufacturing. Wuhan has accelerated innovation by facilitating the translation of research into useful applications and bridging the gap between academia and industry [1]. Cities around the world can adopt this model by promoting cooperation between academic institutions, research facilities, and commercial enterprises [15]. This will guarantee that technological advancements are closely correlated with the demands of the market and societal issues.

Another essential component of Wuhan's urban development strategy has been fostering an environment that is business-friendly [14]. The city has simplified its regulations and provided incentives to draw in both foreign and domestic businesses. This has increased economic growth and strengthened the city's industrial base, which has helped it withstand shocks to the economy. Wuhan's strategy can be applied to other cities by streamlining bureaucratic procedures, providing tax breaks, and establishing an investment-friendly regulatory environment. Attracting the private sector investment required to support urban development requires a friendly business environment.

Wuhan's creation of high-tech zones is especially notable because it prioritizes both economic expansion and well-designed living areas [13]. These zones are intended to be integrated urban areas where people can live, work, and play, not just industrial hubs. This all-encompassing method of urban planning ensures that liveability is not sacrificed for economic growth. Wuhan serves as a model for cities planning to create high-tech zones. These zones should be planned with the needs of the local population, offering easy access to housing, healthcare, education, and recreational resources.

The necessity to remain focused on sustainable development while being ready to tackle unforeseen obstacles is one of the most significant lessons to be learned from Wuhan's experience. Wuhan's response to environmental and economic crises has shown how crucial resilience is to urban design. The city has demonstrated that it is possible to pursue long-term objectives of sustainability and growth while simultaneously making adjustments to changing conditions. In order to respond to obstacles without sacrificing their overarching future vision, cities must incorporate flexibility into their development plans.

Last, but not least, Wuhan's emphasis on incorporating locals in the development process has been essential to building a sense of community and civic pride among the populace. Wuhan's decision-making process has involved its citizens, strengthened the communities and increased their commitment to the city's prosperity [2]. This strategy emphasizes the value of resident voices being heard and needs meeting through participatory planning in urban development. Other cities can take a cue from this by implementing inclusive development strategies that empower residents and produce more responsive and equitable urban environments.

In conclusion, Wuhan's urban development provides a thorough model for other cities. Wuhan's experience offers important lessons on creating a thriving, sustainable city, from encouraging innovation and creating business-friendly environments to ensuring resilience in the face of challenges and balancing economic growth with quality of life. Cities worldwide can successfully negotiate contemporary urban development challenges and create liveable and economically vibrant environments by implementing these strategies.

References

1. Appelbaum RP, Cao C, Han X, Parker R, Simon D (2018) Innovation in China: challenging the global science and technology system. John Wiley & Sons, New York
2. Chang Y (2010) An application of sustainable livelihoods approach to a housing related study in urban China: the case of Shanghai Lane, Wuhan. Doctoral dissertation, Newcastle University
3. Chen X, Li B (2022) Rethinking cultural creativity and tourism resilience in the post-pandemic era in Chinese traditional villages. Sustainability 14(19):12371
4. Cheshmehzangi A (2016) China's new-type urbanisation plan (NUP) and the foreseeing challenges for decarbonization of cities: a review. Energy Procedia 104:146–152
5. Cheshmehzangi A (2023) Regenerating local industries and urban experiences that matter. In: Mapping urban regeneration: city life experiences in Yunnan, China. Springer Nature Singapore, Singapore, pp 23–75
6. Cheshmehzangi A, Tang T (2022) Yangtze Mid-River Delta: China's strategic fulcrum of dynamic growth pole in the form of middle triangle. In China's city cluster development in the race to carbon neutrality. Springer Nature Singapore, Singapore, pp 93–105
7. Cheshmehzangi A, Tang T (2022) China's city cluster development in the race to carbon neutrality. Springer, Singapore
8. Chien SS, Woodworth M (2022) Entrepreneurial and modular urbanism in China: New cities and new areas in the 2000s. Trans Plann Urban Res 1(3–4):219–234
9. Finance Competitiveness (2020) China's doing business success: drivers of reforms and opportunities for the future. World Bank, Washington, DC
10. Gai Q, Li Z, Hu H (2022) Strategies for China's historic districts regeneration in responding to public health emergencies. Sustainability 14(21):14020
11. Huff AG (2022) The truth about Wuhan: how i uncovered the biggest lie in history. Simon and Schuster, New York
12. Kennedy S (2020) China's uneven high-tech drive. Center for Strategic and International Studies (CSIS). Available from: http://csis-website-prod.s3.amazonaws.com/s3fs-public/publication/200302_Kennedy_ChinaUnevenDrive_v3.pdf
13. Liu S (2023) Regeneration strategies for "Anchor+" innovation district: a case study of the Tianhe Changban Plot. Doctoral dissertation, Politecnico di Torino
14. Pan C, Wu S, Li E, Li H, Liu X (2023) Identification of urban functional zones in Macau Peninsula based on POI data and remote information sensors technology for sustainable development. Phys Chem Earth, Parts a/b/c 131:103447
15. Raven R, Sengers F, Spaeth P, Xie L, Cheshmehzangi A, De Jong M (2019) Urban experimentation and institutional arrangements. Eur Plan Stud 27(2):258–281
16. WEHDZ (n.d.) Entering optics valley. Wuhan East Lake New Technology Development Zone Government Affairs Network <<走进光谷. 武汉东湖新技术开发区政务网>>. Available from: https://www.wehdz.gov.cn/
17. Wuhan Government (2021) Wuhan City's 14th Five-Year Plan for National Economic and Social Development <<武汉市国民经济和社会发展第十四个五年规划和 2035 年远景目标纲要. 武汉市人民政府>>. Available from: https://zrzyhgh.wuhan.gov.cn/zwgk_18/fdzdgk/ghjh/zzqgh/202112/P020211202640113900373.pdf
18. Wuhan Government (2022) A summary of Wuhan's development and reform in the past ten years <<数说武汉发展改革这十年>>. In Changjiang Daily, Available from: https://www.wuhan.gov.cn/ztzl/22zt/xszcbgh/sfgw/202210/t20221019_2061860.shtml
19. Wuhan Municipal People's Government (2024) Wuhan City's "Government Work Report" in 2024. <<2024年武汉市《政府工作报告》>>. Available from: https://www.wuhan.gov.cn/sy/whyw/202402/t20240209_2359248.shtml

20. Xie K, Song Y, Zhang W, Hao J, Liu Z, Chen Y (2018) Technological entrepreneurship in science parks: a case study of Wuhan Donghu High-Tech Zone. Technol Forecast Soc Chang 135:156–168
21. Xu J, Zou G, Chung CKL (2022) (Re)shaping urban governance through state-business interaction in inland China's emerging industries. China Review 22(4):77–104
22. Zhang S, Zhao J, Jiang Y, Cheshmehzangi A, Zhou W (2023) Assessing the rural–urban transition of China during 1980–2020 from a coordination perspective. Land 12(6):1175

Chapter 10
Tianjin: From Coastal Advantages to Large-Scale Urban Transformation

1 Introduction to Tianjin Municipality

With a rich historical background, Tianjin has become a northern Chinese city of lasting significance. Tianjin, which is situated by the Bohai Sea, has long been an important entry point to Beijing, the capital of the country [10, 19]. Its significance as a major port and trading hub during the imperial era, when it served as an essential conduit for goods and cultural exchange between China and the wider world [13], emphasizes its historical significance.

In the modern era, Tianjin is a thriving city that is an important hub for transportation and the economy. Due to its advantageous location, which promotes industrial growth and eases international trade [15, 27], it is now a major player in China's economic landscape [2]. As one of China's four direct-controlled municipalities, Tianjin's significance is further highlighted by this fact, which highlights the city's distinct administrative and economic position [16]. Due to its high degree of political and economic autonomy, the city is able to carry out ambitious development plans that meet local needs as well as national objectives.

Tianjin has initiated a number of all-encompassing development plans in recent years with the goal of strengthening its position as a major regional and global economic force. The city's dedication to technological innovation and industrial upgrading is at the heart of these plans. The growth of high-tech industries has been Tianjin's top priority, especially in fields like aerospace, biomedicine, and new energy vehicles. Extensive R&D initiatives aimed at positioning Tianjin as a leader in China's transition to an innovation-driven economy are in place to support these industries.

Sustainable environmental practices are also highly prioritized in Tianjin's development plans. The city is aware of the difficulties brought about by the fast urbanization and industrialization of the region, especially with regard to resource depletion and environmental degradation [3]. Tianjin has responded by putting in place a number of laws meant to encourage green growth. These consist of the creation of

A. Cheshmehzangi and T. Tang, *30 Years of Urban Change in China's 10 Core Cities*, Urban Sustainability, https://doi.org/10.1007/978-981-97-8846-0_10

eco-friendly industrial parks [4, 21] the enlargement of green areas, and the encouragement of clean energy. Sino-Singapore Tianjin Eco-city is a flagship project that aims to establish a model for sustainable urban living in China [9, 26], and it is one of the most prominent examples of Tianjin's commitment to sustainable development.

Furthermore, Tianjin's development strategy remains centered on its status as a significant port city. One of the biggest ports in the world, the Port of Tianjin, is a vital entry point for global trade. Intending to make Tianjin a major hub for international logistics, the city has made significant investments in modernizing and enlarging its port infrastructure [25]. Establishing the Tianjin Free Trade Zone, which attempts to draw in foreign investment and strengthen the city's capacity for international trade, complements these initiatives.

Tianjin's development plans prioritize improving the quality of life for its citizens in addition to its economic goals. The city has implemented multiple urban renewal initiatives to enhance public services, protect its cultural legacy, and upgrade infrastructure. These initiatives demonstrate Tianjin's dedication to developing a liveable, aesthetically pleasing urban setting that balances modernity and tradition. Tianjin's skyline, a harmonious fusion of modern skyscrapers and historical architecture, represents the city's capacity to combine modern urban development with its rich cultural legacy.

The significance of regional integration is also emphasized in Tianjin's development plans. One of China's most significant economic corridors, the Beijing-Tianjin-Hebei (Jing-Jin-Ji) region, includes the city as a major component [5]. In order to promote regional economic growth and address common issues like pollution and traffic congestion, Tianjin aims to deepen its connections with Beijing and Hebei Province through improved connectivity [6] and coordinated development initiatives.

By 2035, according to its development plans, Tianjin wants to have established itself as one of the world's most developed cities. The city hopes to establish itself as a hub for global trade, an innovation hub, and an example of sustainable urban growth. To meet these objectives, it will take sustained investment in technology, education, and infrastructure in addition to a dedication to social justice and environmental sustainability.

Tianjin is a tribute to China's rich historical and contemporary tapestry because of its deep historical roots, economic prominence, and advantageous location. The ambitious and forward-thinking vision of progress embodied in the city's development plans represents the nation's ongoing journey toward prosperity and global influence. As Tianjin develops further, it will surely become a role model for other Chinese cities and cities worldwide, showing how to combine modernity and tradition, economic expansion and sustainability, and local development and international involvement (see Figs. 1, 2, 3, 4, and 5).

Fig. 1 Tianjin's Development in 1984 (map is extracted from open and accessible maps on Google Earth, generated in August 2024)

Fig. 2 Tianjin's Development in 1990 (map is extracted from open and accessible maps on Google Earth, generated in August 2024)

1.1 Tianjin's Main Development Over the Past 30 Years

Tianjin has changed dramatically over the last three decades, going from an industrial center to a bustling, contemporary city. The city has advanced its position in the coordinated development of the Beijing-Tianjin-Hebei (Jing-Jin-Ji) region over the last ten years. By concentrating on its role as "one base and three regions"—a foundation of advanced manufacturing and three important areas: high-tech industries, contemporary services, and a thriving port economy—this strategic shift has accelerated the conversion of its strategic potential into concrete development momentum.

Tianjin has demonstrated adeptness in assimilating into the wider national strategy by seizing opportunities with a proactive mindset. Tianjin has improved regional

Fig. 3 Tianjin's Development in 2000 (map is extracted from open and accessible maps on Google Earth, generated in August 2024)

Fig. 4 Tianjin's Development in 2010 (map is extracted from open and accessible maps on Google Earth, generated in August 2024)

cooperation and interaction by encouraging Beijing and Hebei to develop in different but complementary ways. Thanks to this strategy, Tianjin has been able to draw in a large number of top-notch businesses, resources for innovation, and large-scale projects, especially those involving the transfer of non-capital functions from Beijing. One excellent example is the Binhai-Zhongguancun Science Park, where more than 3,200 businesses have registered, demonstrating Tianjin's expanding status as a center for innovation and technology [18].

Furthermore, Tianjin has been instrumental in advancing the Xiongan New Area, a national project to build a new metropolis that will relieve pressure on Beijing. The city has given this developing area full support by utilizing its first-rate resources in urban planning, education, and port logistics. A pillar of Tianjin's economy, the

Fig. 5 Tianjin's Development in 2024 (map is extracted from open and accessible maps on Google Earth, generated in August 2024)

port has undergone significant renovations to meet the demands of contemporary international trade. Tianjin Port has grown into an important hub for the Beijing-Tianjin-Hebei region and a vital platform for China's Belt and Road Initiative as a result of President Xi Jinping's directive to build a world-class smart and green port [18]. Tianjin's dedication to sustainability and innovation is demonstrated by the port's conversion into a state-of-the-art, zero-carbon facility that includes the world's first intelligent zero-carbon terminal and an autonomous driving demonstration zone. In 2021, the port's throughput of containers exceeded 20 million TEUs, solidifying its position as a hub for international shipping in northern China.

Transportation infrastructure has also seen notable advancements with the construction of a high-density road and rail network that effectively links Tianjin with Beijing and other major cities in the region. The Beijing-Tianjin and Beijing-Binhai lines are two of the five high-speed rail corridors that have significantly shortened travel times [18]. This has improved the one-hour transportation circle that connects the major cities in the Jing-Jin-Ji region. More social and cultural interactions as well as economic integration between the cities have been encouraged by this increased connectivity.

The "Jin City" and "Bin City" twin-city development pattern in Tianjin illustrates a change in the city's growth model from outward expansion to inward upgrading. This new strategy places a strong emphasis on developing sustainable development, boosting infrastructure, and improving urban life quality [18]. The emphasis now is on developing an urban environment that is balanced and promotes both economic growth and a high standard of living for its citizens.

Tianjin's development plans are moving more quickly as it stays a part of the building of the world-class Beijing-Tianjin-Hebei city cluster. The city is redefining itself as a leader in innovation, sustainable development, and urban planning in addition to growing as a major force in the region's economy. Through these initiatives, Tianjin hopes to set an example for other Chinese cities attempting to balance the demands of growing urbanization with the preservation of their natural environment and economic vitality.

With the future in sight, Tianjin's trajectory is expected further to improve its standing on the domestic and global arenas. Its innovative, collaborative, and sustainable growth-oriented strategic initiatives are expected to remain a model for other Chinese cities and other cities worldwide [18]. The city will continue to play a significant role in China's larger regional and economic development plans as it changes, highlighting the potential of a well-thought-out, proactive approach to urban planning.

2 Tianjin and Its Main Development Strategies

The following are the main objectives of the economic and social development in the next five years: they are all centered around the realization of the visionary goal of essentially creating a modern socialist metropolis by 2035; they take into account all of the opportunities, advantages, and conditions of Tianjin's development; and they adhere to the unity of goal-oriented, problem-oriented, and result-oriented.

The fundamental implementation of the functional positioning of "one base and three regions" is a crucial tactic. A number of internationally competitive industrial clusters are formed, the proportion of strategic emerging industries has increased significantly, the national advanced manufacturing R&D base is essentially completed, and the independent and controllable, safe, and efficient industrial chain is more complete [22]. The northern region has gained prominence as a hub for international shipping, and significant progress has been made in the development of smart and green ports. A healthier and more benign financial ecosystem has resulted from a significant improvement in the ability and level of financial services to support the real economy, prevent and control financial risks, and deepen financial reform. A higher level of open economy will essentially form a new system, market participants will be more dynamic, and the business environment will be at the forefront of the nation. Reform and opening up will require new steps, improvements to the system and mechanism adapted to the new development concept and the requirements of high-quality development.

Originally, "Jin City" and "Bin City" formed a twin-city development pattern. Numerous modern service marking districts will be established, and "Jin City's" modern service function will be greatly enhanced. The "Bin Cheng" city's overall supporting capacity has been greatly increased, and a stunning new city by the coast has essentially been built with an ecologically sound, intelligent port-product-city integration that is pleasant to live and work in [22]. Additionally, the spatial layout has been more optimally designed with a sensible division of labour, complementary functions, synergy, and efficiency.

Making sure that excellent economic development is elevated is the alternative strategy. The new development concept has been thoroughly and thoroughly implemented, and based on a notable improvement in quality and efficiency, the economy has been able to develop sustainably and healthily [22]. An important source of China's independent innovation and a major source of original innovation has been built thanks to the significant strengthening of the innovation capacity and the expansion of innovative enterprise clusters. The service economy system will be further refined following the status of a modern metropolis, the industrial structure will be more optimized, and the construction of an international consumption center city and a regional trade center city will come first, forming a higher level of dynamic balance between supply-driven demand and supply-created demand.

The degree of urban civilization has also advanced, which is an important factor. The quality of people's ideology and morality, the quality of science and culture, the quality of their physical and mental health, the soundness of the public cultural service system and the cultural industry system, the enrichment of spiritual and cultural life, and the significant strengthening of cultural soft power are all expected outcomes of the deeper ingraining of the core socialist values in people's hearts [22].

Furthermore, the achievement of new results in the building of an ecological civilization is significant for the city. The development and protection of territorial space has followed an optimized pattern; the "871" major ecological construction projects have advanced significantly; the green transformation of production and lifestyle has produced impressive outcomes; the allocation of energy resources has become more rational; the efficiency of utilization has greatly improved [22],the total amount of major pollutants discharged has continued to decline; the ecological environment has significantly improved; and living environments, both urban and rural, have become greener and more liveable.

The well-being of individuals has advanced to a new degree, which is another significant development. This entails creating more full-time and higher-quality jobs, essentially coordinating the increase in residents' income and economic growth and speeding up the failures of social projects. This factor helps in raising the standard of equalization for basic public services, greatly boosting the standard of education modernization, enhancing the multi-level social security system, enhancing the health system, and broadly advancing the rural revitalization strategy.

The recent improvement in governance effectiveness is a significant development. There will be greater perfection in socialist democracy and the rule of law, greater social justice [1], better government performance, and much increased administrative credibility and efficiency [14]. Development security has been more firmly guaranteed, the institutional mechanism for preventing and resolving major risks has become sturdier, the emergency response capacity for public emergencies has been significantly strengthened [22], and the level of urban management refinement and social governance has improved significantly.

2.1 Tianjin's Binhai New Area: A New Development Zone

The Binhai New Area is a remarkable illustration of Tianjin's extensive urban development during the previous three decades. The Binhai New Area was envisioned as a major development zone in 2006 with the goal of converting an underdeveloped coastal area into a thriving urban and economic hub [17]. The region has experienced incredible changes over time, developing into a vibrant center that is essential to both China's larger regional development plans and the economy of Tianjin (see Figs. 6, 7, 8, and 9).

The development of Binhai New Area into a contemporary urban and economic hub is evidence of the success of China's urbanization and strategic planning initiatives. The area was planned from the beginning to take advantage of the region's natural and infrastructural advantages as well as its strategic location next to the Beijing-Tianjin-Hebei (Jing-Jin-Ji) region along the Bohai Sea [17]. The goal of the government's creation of the zone was to make it not only a vital component of

Fig. 6 The Binhai New Area in 1985 (map is extracted from open and accessible maps on Google Earth, generated in August 2024)

Fig. 7 The Binhai New Area in 2006 at earlier stages of development (map is extracted from open and accessible maps on Google Earth, generated in August 2024)

Fig. 8 The Binhai New Area in 2012 (map is extracted from open and accessible maps on Google Earth, generated in August 2024)

Fig. 9 The Binhai New Area in 2024 (map is extracted from open and accessible maps on Google Earth, generated in August 2024)

Tianjin's economic development but also an essential link in China's larger economic system.

By 2020, with the Binhai New Area contributing an astounding 42% of Tianjin's GDP, the area had solidified its position as the city's economic engine. Additionally, the Binhai New Area made a substantial contribution to the city's industrial output: 57% of the regulated industries' total output value originated from this area. The region's fiscal contributions matched its industrial prowess, accounting for 67% of Tianjin's tax revenue at the municipal and district levels.

In addition, the Binhai New Area has drawn trade and investment from abroad, demonstrating its importance in China's international economic interactions. In 2020, the region accounted for 73% of the city's import and export activities in foreign trade, and it drew 80% of Tianjin's actual foreign investment [17]. These numbers demonstrate how the region has succeeded in becoming a hub for global trade and business thanks to a combination of advantageous laws, cutting-edge infrastructure, and advantageous geographic location.

The following year saw the Binhai New Area maintain its strong economic performance. The region's gross regional product grew by 9.4% annually in the first three quarters of 2021, indicating steady economic growth despite uncertainty around the world. The region's regulated industries saw a notable 19.5% increase in total output value, and estimates for the year suggest that the region's economic output may surpass 1 trillion yuan. An additional 17.5% increase in general public budget revenue—tax revenues alone rising by 20.6%—supported this growth trajectory [17].

The Binhai New Area continued to draw strong foreign investment, with $3.34 billion in actual utilized foreign investment, indicating a 15.7% year-over-year increase. The region's economy has grown largely due to this inflow of foreign capital, which has also improved the region's economic connectivity worldwide. Furthermore, the region continued to exhibit strong performance in international trade, with import and export activities totalling $446.58 billion, up 11.2% from the year prior [17]. These numbers highlight the Binhai New Area's significance in the global supply chain and its function as a major hub in China's foreign trade network.

The Binhai New Area has advanced significantly in urban development and modernization in addition to its economic accomplishments. The region has experienced significant infrastructure development, encompassing the erection of contemporary transit systems, residential complexes, and commercial hubs [17]. Urban planning in the area has placed a strong emphasis on integrating sustainable practices and green spaces, demonstrating a dedication to establishing a liveable and eco-friendly urban environment. This strategy has drawn an increasing number of businesses and residents, enhancing the economic and social vibrancy of the region.

Because of its success, the Binhai New Area can serve as an example for other areas of China and the world, showing how large-scale, carefully thought-out urban development projects can stimulate both urbanization and economic growth [17]. The region's explosive growth has also taught us important lessons about how to balance environmental sustainability with economic growth, promote innovation, and draw in talent and capital from around the world.

The Binhai New Area is anticipated to become more and more important in the economic strategies of both the city and the country as Tianjin develops. Tianjin's future will probably be shaped by its continued expansion and development, which will establish the city as a major economic and urban hub in China as well as an important actor in the world economy [17]. The success story of the Binhai New Area serves as a reminder of how crucial strong infrastructure, a strategic vision, and a dedication to sustainable development are to long-term urban and economic prosperity.

3 Tianjin's Projection for Future Development

The city's aspirational goals and flexible approaches to sustainable development are reflected in Tianjin's current and upcoming development plans, which highlight the city's strategic importance in China's economic environment. Tianjin has developed into a major economic center in the last 30 years, and the Binhai New Area has been essential to this development [17]. Since its establishment in 2006, the Binhai New Area has grown quickly into a sophisticated urban hub that significantly boosts Tianjin's economic indicators [20]. This region alone was responsible for 42% of the city's GDP by 2020, demonstrating its prominence as a hub for investment, international trade, and industrial output [23].

The city of Tianjin's future development plans are designed to build on these achievements, emphasizing high-quality, sustainable growth. The city's 2024 GDP growth target of roughly 4.5% reflects a cautious but optimistic approach to economic expansion. This target signals a commitment to raising living standards in tandem with economic performance [23]. It is accompanied by goals to ensure that general public budget revenue and per capita disposable income grow in tandem with the overall economy.

The improvement of fixed asset investment is one of Tianjin's plans for future growth and is predicted to see positive growth [23]. This is essential for meeting infrastructure needs and maintaining long-term economic momentum. In keeping with its goal of leveraging domestic consumption as a growth engine, the city also plans to raise total retail sales of consumer goods by 6%. Another important component of Tianjin's development plans is job creation, with 350,000 new jobs in urban areas being the goal [12, 24]. This is a component of a larger plan to keep urban unemployment at a low level, about 5.5%, necessary for social stability and economic resilience [23]. Additionally, in order to preserve purchasing power and consumer confidence, the city intends to limit increases in consumer prices to roughly 3%. This balances economic growth with the management of inflation.

Tianjin's environmental objectives are closely linked to its development. The city is dedicated to meeting federal goals for pollution reduction, energy conservation, and emission reduction because these goals are essential to its larger goal of promoting sustainable urban growth [23]. This indicates an increasing understanding of the need to strike a balance between environmental protection and economic growth in order to prevent Tianjin's growth from jeopardizing the city's ecological stability.

Apart from pursuing economic and environmental goals, Tianjin is also concentrated on augmenting its involvement in global trade and cross-border commerce. The city has achieved great success in international trade; in the first half of 2024 alone, total imports and exports exceeded $55 billion. The city intends to keep growing its maritime economy by fusing ports, businesses, and cities to build a network of trade that is more effective and adds value [23]. Tianjin's aspirations to become a global commerce leader are demonstrated by initiatives like the creation of a national innovation demonstration zone for import trade promotion and the growth of cross-border e-commerce.

In addition, the city's future plans place a strong emphasis on the development of smart ports and technology innovation with the goal of making Tianjin a global center for logistics and trade. One example of this is the "single window" service's expansion along the entire trade chain, which improves port efficiency and expedites customs procedures. Tianjin's development plans for the upcoming years, taken as a whole, show a balanced approach to technological innovation, environmental sustainability, and economic growth [23]. By concentrating on these important areas, Tianjin is establishing itself as a prominent city with a promising future for continued prosperity and advancement—not just in China but also internationally.

3.1 Tianjin and Its Enduring Challenges in Development

Despite its rapid development, Tianjin, one of China's most important industrial and commercial hubs, faces many obstacles. The city has come a long way in the last few decades, but industrial restructuring is still a difficult and continuous undertaking. The city faces difficulties in shifting to more inventive and sustainable industries since the legacy industries—especially those in heavy manufacturing and petrochemicals—continue to rule the economy.

The relatively small "chassis" or foundation of new kinetic energy industries is one of the most critical issues. Although Tianjin has made progress in creating high-tech zones and encouraging innovation, these industries are still insufficiently developed to propel the city's economic transformation completely. The sluggish pace of industrial restructuring in China indicates a larger national trend in which traditional industries are ingrained deeply and are challenging to transform rapidly. The task facing Tianjin is to steadily reduce its reliance on older, more polluting industries while accelerating the growth of emerging industries like biotechnology, information technology, and green energy.

Significant improvements are also needed to ensure Tianjin's private economy remains vibrant. The city has enacted some policies, such as tax breaks and streamlined regulations, to encourage the growth of the private sector, but the results have been uneven. Numerous private businesses still struggle to get funding through bureaucratic procedures and compete with state-owned enterprises (SOEs), which frequently get special treatment [23]. A thriving private economy is necessary for innovation, job creation, and sustainable growth, so this lack of dynamism in the private sector significantly hampers Tianjin's wider economic goals.

Furthermore, although Tianjin's business climate has improved, it still falls short of the cutting-edge standards observed in other prestigious Chinese cities like Shanghai and Shenzhen. Inadequate infrastructure, unclear regulations, and ineffective administration are common problems businesses in Tianjin face. These problems may discourage foreign as well as domestic investors, reducing the city's ability to draw talent and high-caliber investments. It is necessary to step up the local government's continuous efforts to improve the business climate by cutting red tape, boosting customer service, and making doing business easier overall.

Meeting the changing needs of its citizens for a high quality of life is another crucial challenge for Tianjin. Expectations for public services, healthcare, education, and environmental quality are growing as the city expands and becomes more modern. But there's still a disconnect between these expectations and what's happening [23]. For instance, many locals continue to have serious concerns about pollution despite advancements in air and water quality. Comparably, there are differences in access to high-quality healthcare and education services between urban and rural areas as well as between city districts.

In addition, the city must take precautions against risks and hidden hazards in strategic locations. These include financial risks associated with excessive debt, especially for state-owned enterprises (SOEs) and local governments, which, if not properly managed, may threaten economic stability. Concerns exist regarding the safety and robustness of the city's infrastructure, especially in light of the increasing frequency and severity of natural disasters like flooding brought on by climate change [23].

And lastly, there are issues with the government structure itself. The new circumstances and the demands of contemporary governance may be difficult for some officials and employees to adjust to. This entails having the capacity to put creative policies into action, handle challenging urban problems, and adapt to the demands of a society that is changing quickly. To effectively navigate these challenges and maintain its development trajectory, Tianjin must improve the responsiveness and capacity of the local government. In conclusion, Tianjin has come a long way in its development, but to maintain this progress and raise the standard of living for its citizens, the city needs to confront these issues head-on. To guarantee that Tianjin continues to be a vibrant and resilient urban center in the coming years [23], a coordinated effort involving all governmental, industrial, and social levels will be necessary.

4 Key Lessons Learnt from Tianjin for Other Cities

The Bohai Sea city of Tianjin has made the most of its coastal location to establish itself as a major hub for trade and commerce in northern China [11]. Tianjin's economic vitality has historically been attributed to its strategic location as a gateway to Beijing. Today, this advantage continues to underpin Tianjin's growth, particularly in international trade and maritime industries. The city's port, one of the largest in the world, plays a crucial role in facilitating this economic dynamism, connecting China's interior to global markets.

Combining modern development with historical charm is one of Tianjin's noteworthy accomplishments [8]. The city's rich history is reflected in its notable cultural landmarks, which include Ancient Culture Street and the Italian-style concession area. Tianjin is distinguished from other Chinese cities by its distinctive architectural fusion, which is achieved through the coexistence of these historical sites with contemporary skyscrapers and urban developments [7]. The city's attractiveness as a modern metropolis is increased by this tasteful blending of the old and the new, which also protects the city's cultural legacy.

The Tianjin Economic-Technological Development Area (TEDA), a high-tech industrial zone that has turned into a model for economic resilience, has greatly aided Tianjin's development. TEDA is a key component of Tianjin's economic strategy due to its success in luring and keeping high-tech industries. The region has developed into a center for innovation, luring businesses from abroad and domestically to establish themselves in Tianjin. The city's economy has become more diversified and more resistant to changes, due to TEDA's emphasis on high-tech industries.

Another important component of Tianjin's urban design is sustainability. The Sino-Singapore Tianjin Eco-City is a trailblazing project in sustainable development that exemplifies how cross-border cooperation can produce creative answers for urban living (see Figs. 10, 11, 12, and 13). This eco-city was created in collaboration with Singapore and is intended to serve as a model for sustainable urbanization. It uses green practices and technologies to make the city more liveable and ecologically friendly. The accomplishment of this project serves as evidence of the value of global collaboration in tackling the issues associated with contemporary urban development.

Because it is a direct-controlled municipality, Tianjin enjoys administrative autonomy essential for carrying out targeted and efficient development plans. Because of its independence, Tianjin can adapt its policies to the needs and goals of the region, promoting a more dynamic and adaptable style of government. Additionally, it makes it possible for the city to carry out significant development projects like TEDA and the Binhai New Area more successfully and efficiently. Tianjin has effectively incorporated cultural diversity as a valuable asset into its development strategy. Both locals and visitors enjoy a richer urban experience because of the city's diverse neighbourhoods and historical sites, which showcase its rich cultural history. Places such as the Italian-style concession draw tourists and add to the cultural and

Fig. 10 The evolution of the Sino-Singapore Tianjin Eco-City, over marshlands and reclaimed land to a new model of urban planning in China. Map is from 1990s. (maps are extracted from open and accessible maps on Google Earth, generated in August 2024)

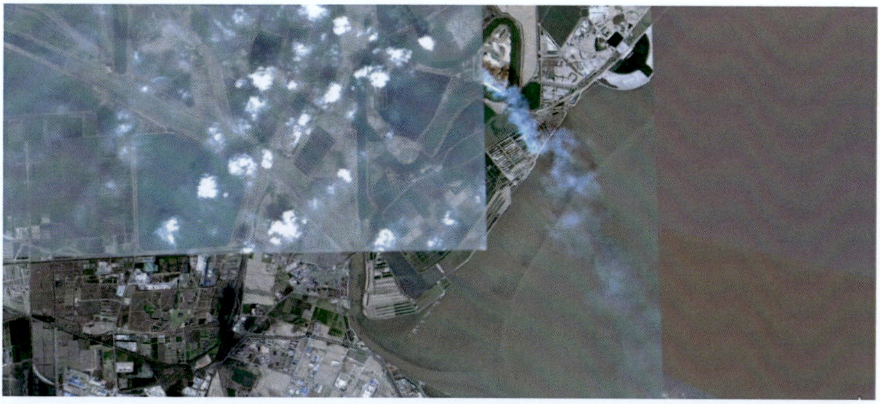

Fig. 11 The evolution of the Sino-Singapore Tianjin Eco-City, over marshlands and reclaimed land to a new model of urban planning in China. Map is from 2000s. (maps are extracted from open and accessible maps on Google Earth, generated in August 2024)

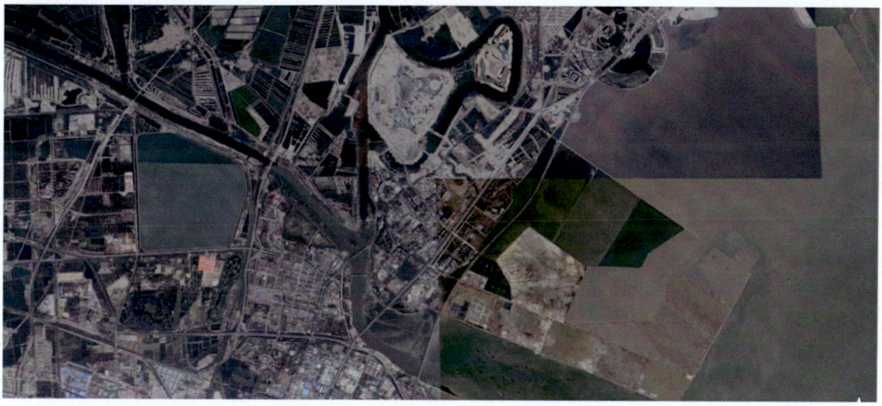

Fig. 12 The evolution of the Sino-Singapore Tianjin Eco-City, over marshlands and reclaimed land to a new model of urban planning in China. Map is from 2010s. (maps are extracted from open and accessible maps on Google Earth, generated in August 2024)

economic life of the city. Tianjin strengthens its image as a city where tradition and modernity coexist by embracing and exhibiting its cultural heritage. In conclusion, Tianjin's growth in recent years serves as an example of how a city can take advantage of its advantageous location, combine old and new, support high-tech industries, embrace sustainability, and profit from cultural diversity.

Fig. 13 The evolution of the Sino-Singapore Tianjin Eco-City, over marshlands and reclaimed land to a new model of urban planning in China. Map is from 2020s. (maps are extracted from open and accessible maps on Google Earth, generated in August 2024)

References

1. Allam Z, Cheshmehzangi A, Jones DS (2023) Climate and social justice in the political landscape of urban resilience and mercantilism: a concluding note. In: Climate and social justice: the political economy of urban resilience and mercantilism. Springer Nature Singapore, Singapore, pp 103–110
2. Chan Y (2020) Economic and strategic significance of development performance of special economic zones from the perspective of maritime industry. Available from: https://commons.wmu.se/cgi/viewcontent.cgi?article=2460&context=all_dissertations
3. Cheshmehzangi A, Dawodu S, Sharifi A (2021) Sustainable urbanism in China. Routledge, New York
4. Cheshmehzangi A, Flynn A, Tan-Mullins M, Xie L, Deng W, Mangi E, Chen W (2021) From eco-urbanism to eco-fusion: an augmented multi-scalar framework in sustainable urbanism. Sustainability 13(4):2373
5. Cheshmehzangi A, Tang T (2022) Beijing-Tianjin-Hebei city cluster: from a political center to a major industrial and innovation chain. In: China's city cluster development in the race to carbon neutrality. Springer Nature Singapore, Singapore, pp 35–46
6. Cheshmehzangi A, Tang T (2022) China's carbon neutrality plan implementation and sustainable development process review. In: China's city cluster development in the race to carbon neutrality. Springer Nature Singapore, Singapore, pp 1–23
7. Cheung PT (2005) Guangzhou and Tianjin: the struggle for development in two Chinese cities. Cities in Post-Mao China. Routledge, New York, pp 17–50
8. Croddy E (2022) Tianjin city. China's provinces and populations: a chronological and geographical survey. Springer International Publishing, Cham, pp 671–679
9. Deng W, Cheshmehzangi A, Deng W, Cheshmehzangi A (2018) Meso level: eco-neighbourhood/community cases in China. In: Eco-development in China: cities, communities and buildings, pp 157–185
10. Dong XC, Wang GX (2012) Coal logistics competency strategies for ports in the Tianjin and Hebei regions around the Bohai Bay in China. Energy Procedia 17:436–443
11. Hao W (2020) Research on the development of Tianjin free trade zone under the background of the Belt and Road Initiative. In: The first interdisciplinary and area studies, p 183
12. Kim H, Allen RM (2018) Glocalizing cures for China's brain drain ills: the thousand talents plan in Shanghai, Tianjin, and Guangdong. Int J Comp Educ Dev 20(1):16–32

13. Kwan MB (1990) The merchant world of Tianjin: society and economy of a Chinese city. Stanford University. Available from: https://www.proquest.com/docview/303850440?fromopenview=true&pq-origsite=gscholar&sourcetype=Dissertations%20&%20Theses
14. Li B, Cao G (2024) Evolution and governance of the structure of marine economic networks in China's coastal areas: based on Sea-related A-share listed companies. Chin Geogr Sci 1–18.
15. Li K (2020) On the road to a modern city: new transportation technology and urban transformation of Tianjin, 1860–1937. Doctoral dissertation, University of Minnesota
16. Liu C, Chen X (2012) The third coming of China's special economic zones: the rise and regional dimensions of Tianjin Binhai new area. In: Rethinking global urbanism. Routledge, pp 127–149
17. NDRC (National Development and Reform Commission) (2021) Focusing on the innovative development of national-level new areas—Tianjin Binhai New Area: building a beautiful "beach city" to promote high-quality development in the new era. <<聚焦国家级新区创新发展—天津滨海新区: 建设美丽"滨城"推动新时代高质量发展>> Available from: https://www.ndrc.gov.cn/fzggw/jgsj/dqs/sjdt/202112/t20211231_1311309.html
18. People's Daily (2022) Ten years of spring and autumn have left an extraordinary mark. Let's take a look at the new changes in Tianjin in the past ten years. <<十载春华秋实, 镌刻非凡印记, 一起来看天津10年的新变化>>. Available from: https://www.moj.gov.cn/pub/sfbgw/zwgkztzl/2022zt/20220311fjxzc/20220712ffsn/202207/t20220712_459584.html
19. Qiao Y, Guo Q, Wu X, Wu H, Zhu L, He Y (2021) Environmental risk analysis of surface water based on multi-source data in Tianjin Binhai New Area, China. Environ Monit Assess 193:1–15
20. Shen Q (2017) Planning special economic zones in China. The Routledge companion to planning in the Global South. Routledge, New York, pp 104–114
21. Tan-Mullins M, Cheshmehzangi A, Chien S, Xie L (2017) Smart-eco cities in China: trends and city profiles 2016. Exeter: University of Exeter (SMART-ECO Project). Available from: https://www.academia.edu/download/55600966/Smart-Eco_Cities_in_China_Trends_and_City_Profiles_2016.pdf
22. Tianjin Municipal People's Government (2021) Tianjin's "14th Five-Year Plan" and outline of long-term goals for 2035 <<天津"十四五"规划和二〇三五年远景目标纲要>>. Available from: https://www.ndrc.gov.cn/fggz/fzzlgh/dffzgh/202104/P020210401307524156363.pdf
23. Tianjin Municipal People's Government (2024) Tianjin Municipal Government Work Report 2024. <<天津市2024年政府工作报告>>. Available from: http://tj.people.com.cn/n2/2024/0129/c375366-40730714.html
24. World Bank (2009) Sino-Singapore Tianjin eco-city: a case study of an emerging eco-city in China. World Bank. Available from: https://documents.worldbank.org/curated/en/776301468029076278/pdf/590120WP0P114811REPORT0FINAL1EN1WEB.pdf
25. Wu J (2012) The development of port and the container transport chain: a case study of Tianjin Port. Available from: https://www.diva-portal.org/smash/record.jsf?pid=diva2%3A505012&dswid=9689
26. Xie L, Tan-Mullins M, Cheshmehzangi A (2019) Political ecology of Chinese smart eco-cities. In: Remaking sustainable urbanism: space, scale and governance in the new urban era, pp 57–78
27. Xu H (2022) The road map of the rise of national central cities under the Belt and Road Initiative. In The Belt and Road: industrial and spatial coordinated development. Springer Nature Singapore, Singapore, pp 219–242

Chapter 11
Suzhou: Lessons in Economic Integration, Cultural Preservation, and Green Urbanism

1 A Brief Introduction to Suzhou, China

Suzhou is a city of lasting significance in eastern China because of its colourful past. Over two millennia have seen Suzhou, which is tucked away along the Yangtze River's lower reaches, function as a hub of culture and commerce [15]. Suzhou, long known for its elaborate gardens [12, 27], calm canals, and silk manufacturing, has had a lasting impact on China's cultural legacy as a center of the arts and crafts. Suzhou's classical gardens, which have been recognized as a UNESCO World Heritage site, are a perfect representation of the city's charm [30]. These gardens, which represent the harmony between human ingenuity and nature, are praised for their meticulous design. Examples of these gardens are the Humble Administrator's Garden and the Lingering Garden [28]. Suzhou's canals, which are a relic of the former Grand Canal, add to the city's historical charm by creating a network that symbolizes its importance as a hub for trade and transportation in the past.

Suzhou is a vibrant city that has successfully combined its rich history with contemporary affluence. Its advantageous position as a technological and economic powerhouse in the Yangtze River Delta has helped it rise to prominence [10]. The city's booming high-tech industries and strong manufacturing base highlight its transformation from a historical hub to a modern economic center [14]. Thanks to its ability to draw in multinational companies and promote innovation across a range of industries, Suzhou has emerged as a significant hub in China's economy [6]. The skyline of the city, which contrasts sleek, contemporary architecture with historic neighbourhoods and temples, is a reflection of its rapid urbanization [4, 25]. This combination of the old and the new is more than just a visual change, it's evidence of Suzhou's ability to balance modern innovations with its rich cultural legacy [5, 7, 8]. Suzhou's role as a major player in the global silk industry, maintaining its historical legacy while embracing contemporary technological advancements [11, 22], is another indication of its economic dynamism.

A. Cheshmehzangi and T. Tang, *30 Years of Urban Change in China's 10 Core Cities*, Urban Sustainability, https://doi.org/10.1007/978-981-97-8846-0_11

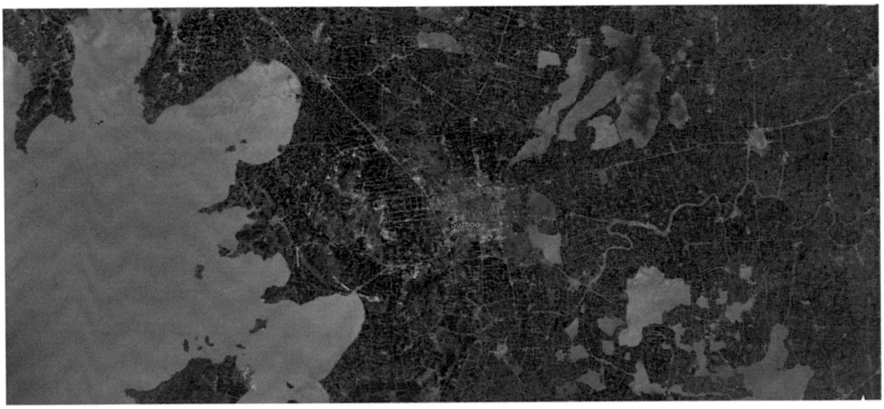

Fig. 1 Suzhou and its regional development in 1984 (Map is extracted from open and accessible maps on Google Earth, generated in August 2024)

Suzhou's urban planning and cultural preservation programs serve as additional evidence of its dedication to maintaining its historical identity in the face of rapid modernization [3, 13]. The city's commitment to a balanced approach to development is demonstrated by its efforts to preserve its historical sites while allowing for growth [9]. Its historic streetscapes have been preserved, and historical details have been incorporated into more recent construction.

Suzhou's significance as a guardian of Chinese heritage is further reinforced by its designation as a nationally famous historical and cultural city [23]. Its dynamic cultural landscape, which includes customary festivals, arts, and cuisine, keeps drawing travellers and academics worldwide [24]. The city's dedication to cultural interchange and education guarantees its rich customs will be carried on to upcoming generations, adding to its allure.

Suzhou is a representation of China's capacity to adroitly combine modern development with its historical legacy as it continues to change [2, 16]. The city represents a singular meeting point of the past and present, resulting in a dynamic and colourful urban environment where modernity and history coexist in perfect harmony [1]. Suzhou is a model for other cities trying to strike a balance between tradition and progress in the twenty-first century because of its remarkable transformation from an ancient hub of trade and the arts to a contemporary economic powerhouse (see Figs. 1, 2, 3, 4, and 5).

2 Suzhou's Main Development in Recent 30 years

The 19th Report on the Competitiveness of Chinese Cities, published by China Social Science Press and the Institute of Financial and Strategic Studies of the Chinese Academy of Social Sciences, ranks Suzhou sixth in China for Comprehensive

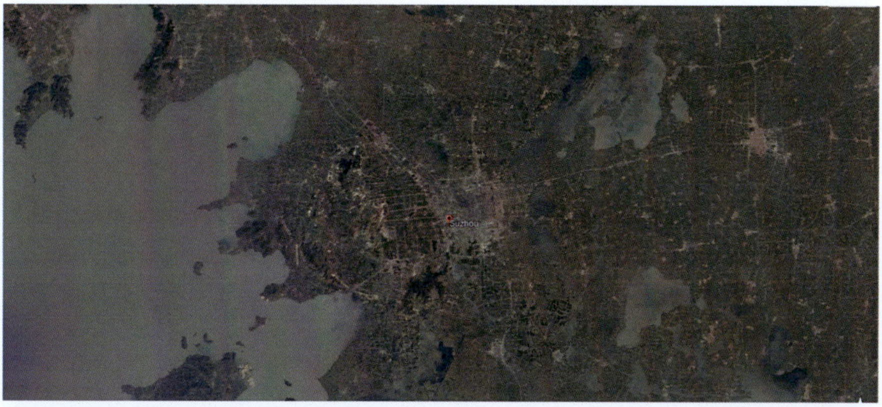

Fig. 2 Suzhou and its regional development in 1990 (Map is extracted from open and accessible maps on Google Earth, generated in August 2024)

Fig. 3 Suzhou and its regional development in 2000 (Map is extracted from open and accessible maps on Google Earth, generated in August 2024)

Economic Competitiveness. Attained in 2021, this ranking highlights Suzhou's noteworthy economic prominence and demonstrates its strategic efforts in conforming to novel development paradigms [19].

The city's unwavering dedication to innovation and economic transformation is responsible for its outstanding performance. Suzhou has continuously prioritized improving new increments and maximizing current assets in order to spur growth. It has actively boosted market vitality through reform and opening-up policies, and it has assiduously increased quality and efficiency through innovation [19]. Suzhou has been able to effectively utilize its assets as a result of its unwavering efforts to accelerate economic dynamism, which has led to notable contributions to the national economy.

Fig. 4 Suzhou and its regional development in 2010 (Map is extracted from open and accessible maps on Google Earth, generated in August 2024)

Fig. 5 Suzhou and its regional development in 2020 (Map is extracted from open and accessible maps on Google Earth, generated in August 2024)

Despite making up only 0.09% of China's total land area, Suzhou has significantly boosted the country's economy, contributing 2.1% of its fiscal revenue, 2% of its GDP, and 6.5% of its total volume of imports and exports. This remarkable economic footprint emphasizes Suzhou's significance as a "ballast stone" for provincial development in addition to highlighting its position as a major economic player [19]. Suzhou has accomplished a significant economic milestone in the last ten years, with its GDP expected to reach 2.27 trillion yuan in 2021, placing it sixth nationally. Furthermore, it was China's first prefecture-level city to earn a GDP of more than two trillion yuan. From 100,000 yuan in 2012 to 177,500 yuan in 2021, the city's per capita GDP increased. The market dynamics in Suzhou are also noteworthy,by the end of 2021, there will be 2.741 million market participants, a threefold increase from the current number [19]. Suzhou is the only province with three major enterprises—Hengli, Shenghong, and Shagang—listed among the world's top 500 companies [19].

Over the past ten years, Suzhou's industrial landscape has seen profound changes. The industrial structure of the city has been improved over time, and in 2016 the tertiary sector overtook the secondary sector for the first time. The industry composition changed in 2021 to 51.3% tertiary, 47.9% secondary, and 0.8% primary sectors, indicating an 8.6 percentage point increase in the value added of the tertiary sector from 2012 to 2021 [19]. This change reflects Suzhou's changing economic profile, which is marked by a more prominent modern economic structure and a strong dual-wheel drive model of secondary and tertiary industries.

A significant development occurred recently when Suzhou announced its strategic focus on the photonics sector at the Photonics Industry Innovation Cluster Development Conference on August 18. By building the Taihu Lake Photonics Center, the city hopes to become a leader in the industry [19]. Suzhou's position as a key player in high-tech industries is reinforced by this initiative, which is part of a larger plan to develop China's photonics industry and establish a world-class innovation cluster (see Figs. 6, 7, 8, and 9 for a mixed-use development example).

Suzhou's regulated industries' combined output value hit a record high of 4 trillion yuan, demonstrating significant industrial growth. With more than 160,000 industrial enterprises and 12,000 high-tech firms spread across 35 industrial categories, the city has also seen impressive growth in its manufacturing sector. The value of industrial output rose by 53.1% by 2021 compared to 2012, averaging 4.8% yearly growth [19]. Significant growth has been observed in the emerging and high-tech industries, which accounted for 54.0% and 52.5% of the total industrial output, respectively.

Furthermore, Suzhou's contemporary service sector has experienced significant expansion, with the added value expected to cross the trillion yuan threshold in 2020

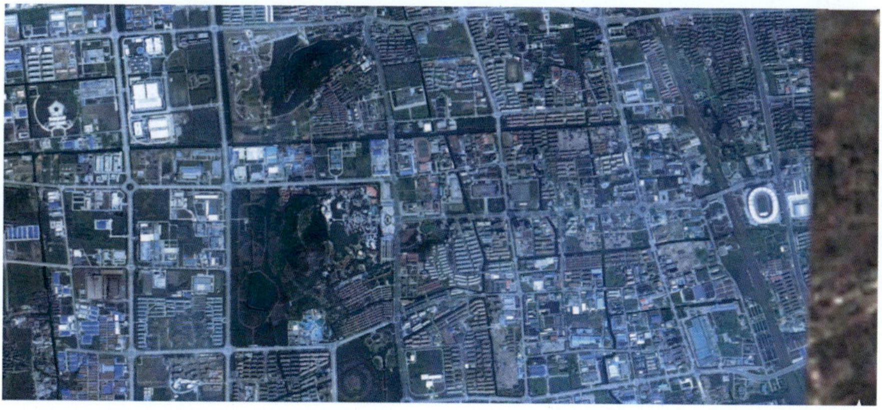

Fig. 6 A mixed-use new urban development zone in Suzhou at its early-to-mid stages of rapid development in 2004 (Map is extracted from open and accessible maps on Google Earth, generated in August 2024)

Fig. 7 A mixed-use new urban development zone in Suzhou in 2009 (Map is extracted from open and accessible maps on Google Earth, generated in August 2024)

Fig. 8 A mixed-use new urban development zone in Suzhou in 2014 (Map is extracted from open and accessible maps on Google Earth, generated in August 2024)

for the first time. The added value increased by 111.1% from 2012 to 1,165.6 billion yuan by 2021, with an average annual growth rate of 8.7%. The emergence of a contemporary service industry system [19], driven by premium services and backed by productive services, is indicative of the changing economic climate in the city. 55% of the value added in the service sector in 2021 came from the productive service sector.

Suzhou is a major economic powerhouse that blends historical significance with cutting-edge development to foster sustained growth and innovation. Its economic achievements and strategic advancements serve as prime examples of this. As part of its economic strategy, Suzhou engages in regional development initiatives and forges solid partnerships within the Yangtze River Delta. Suzhou is now a major

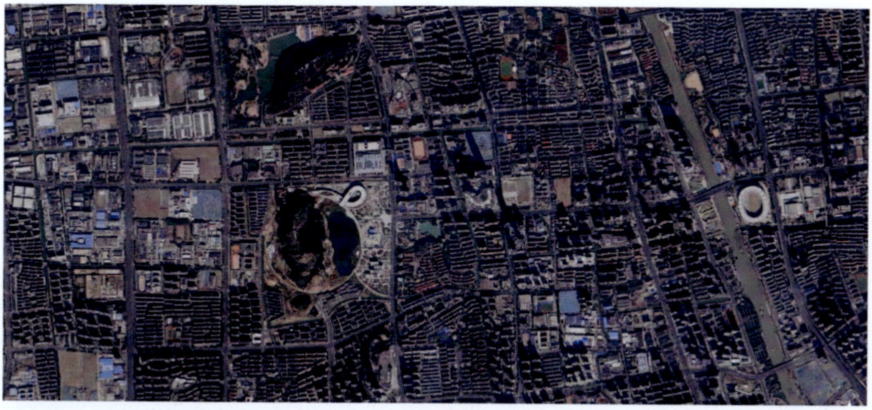

Fig. 9 A mixed-use new urban development zone in Suzhou in 2024 (Map is extracted from open and accessible maps on Google Earth, generated in August 2024)

player in sectors like finance, high-tech innovation, and advanced manufacturing as a result of these initiatives. The city has been effective in drawing in investments and creating a growth-friendly atmosphere, making use of its location to gain from regional collaborations. Its participation in significant infrastructure projects and economic clusters that improve its connectivity and competitiveness on a national and international scale [19] is another aspect of its role in the Yangtze River Delta.

3 A Summary of Suzhou's Development Strategies

By 2035, Suzhou aspires to be a far more advanced city, with a flourishing economy, affluent residents, a beautiful environment, and a highly advanced social civilization. Part of the city's strategic vision is achieving the goals of "striving to be an example, striving to be a model, and leading the forefront." It also hopes to establish a reputation as a center for humanism, innovation, openness, ecology, liveability, and responsible governance. Suzhou wants to develop into a highly advanced socialist metropolis, an important hub for national history and culture, a well-liked vacation spot, and an important player in the Yangtze River Delta. This project will contribute to the development of an elite urban cluster and serve as a showcase for the new paradigm of "Strong, Rich, Beautiful, and High."

By 2035, Suzhou's per capita GDP is predicted to have doubled from 2020 levels, and the city's economic growth will be accompanied by an increase in the income of its citizens. The city is expected to reach consumption and per capita income levels comparable to developed countries. Suzhou's comprehensive economic strength, which enhances its soft power and core competitiveness, will make it one of the world's leading cities [18]. With more control over key technologies, a bigger part in global supply, value, and industrial chains, and continued leadership in national innovation metrics, its economic system will be modernized.

The city will also encourage a free market economy that is more advanced and stable, serving as a national model for strong economic integration with globalization. Urban and rural development [29], regional cooperation, Suzhou's enhanced service-driven capability, and agricultural modernization will set national standards. The city's commitment to raising living standards is reflected in the expanding middle class, increasing wealth, and overall human development. Basic public services will be available to all, with modern education serving as a model. Green and low-carbon development will be the main focus of environmental improvements, with Suzhou to become a model city for "Beautiful China." Suzhou's transformation into a "strong cultural city" will see its social civilization reach unprecedented heights [18]. Suzhou will become a national governance capability leader with the modernization of most governance systems.

3.1 Suzhou Industrial Park (SIP): One of China's Earliest Success Stories in Sustainable Development

Situated in Suzhou City on the eastern side of Jiangsu Province, the Suzhou Industrial Park (SIP) is a bright example of international cooperation and economic development. 278 square kilometers make up SIP, which was established in February 1994 and formally inaugurated in May of the same year. Of that area, 80 square kilometers are dedicated to the Sino-Singapore Cooperation Zone (Suzhou Industrial Park Management Committee, n.d.). Being a significant joint venture between the governments of China and Singapore, this park serves as a major platform for China's reform and opening-up initiatives. Generally, the project is known and acknowledged as a "window for China's engagement with the global economy" and a "successful example of international cooperation" (Suzhou Industrial Park Management Committee, n.d.).

In 2023, SIP achieved noteworthy economic benchmarks. The park's gross domestic product (GDP) rose to 368.6 billion yuan in the previous year, a 5.9% increase. The public budget's revenue grew by 6.1% as well, totalling 41.11 billion yuan. The total output value of the park's regulated industries increased by 0.1% to 650.94 billion yuan, while investments in fixed assets increased by 25.3% to 59.29 billion yuan. Consumer goods retail sales saw a noteworthy 26.1% increase to 117.31 billion yuan, while industrial investment increased by 26.1% to 22.64 billion yuan. The volume of imports and exports came to a total of 606.97 billion yuan, while the overall retail value increased 6.9% to 117.31 billion yuan (Suzhou Industrial Park Management Committee, n.d.). Additionally, foreign investment totalling 1.951 billion USD was drawn to the park.

The remarkable outcomes of SIP validate its standing as one of China's most excellent, transparent, and productive development zones. It has consistently received awards for its innovative and business-friendly environment, and from 2016 to 2023, it topped the comprehensive assessment of national economic and development zones eight times in a row. The Ministry of Science and Technology recognized the park in 2018 for its efforts to establish a premier technological park. The park was designated as an Advanced Collective on the 40th Anniversary of Reform and Opening-up in Jiangsu Province (Suzhou Industrial Park Management Committee, n.d.). SIP has made a name for itself as an effective model of industrial and economic growth. Its achievements also demonstrate how important a role it plays in fostering international cooperation and economic expansion (see Figs. 10, 11, 12, 13, and 14).

Fig. 10 The unique urban development of Suzhou Industrial Park (SIP) from 1985 to 2024. Map is from 1985. (Maps are extracted from open and accessible maps on Google Earth, generated in August 2024)

Fig. 11 The unique urban development of Suzhou Industrial Park (SIP) from 1985 to 2024. Map is from 1990s. (Maps are extracted from open and accessible maps on Google Earth, generated in August 2024)

Fig. 12 The unique urban development of Suzhou Industrial Park (SIP) from 1985 to 2024. Map is from 2000s. (Maps are extracted from open and accessible maps on Google Earth, generated in August 2024)

Fig. 13 The unique urban development of Suzhou Industrial Park (SIP) from 1985 to 2024. Map is from 2015. (Maps are extracted from open and accessible maps on Google Earth, generated in August 2024)

4 Suzhou's Projection for Future Development

The city aims to make significant progress in several areas and has set ambitious yet clear goals for its future social and economic development. The main goals include a regional GDP growth rate exceeding 5% and a corresponding increase of approximately 4% in general public budget revenue. It is projected that overall retail sales of consumer goods will rise by roughly 5.5%, while fixed asset investment will rise by about 5% [20]. To ensure that the manufacturing value-added proportion of the GDP does not fluctuate, steps will be taken. Along with maintaining stable import

Fig. 14 The unique urban development of Suzhou Industrial Park (SIP) from 1985 to 2024. Map is from 2024. (Maps are extracted from open and accessible maps on Google Earth, generated in August 2024)

and export ratios, the city also wants to maintain stable real use of foreign capital relative to the country and province.

The city's commitment to research and development states that approximately 4.15% of GDP will be allocated to costs related to research and experimental development. The expectation is that per capita disposable income will increase in line with the growth of the economy as a whole, with consumer prices expected to rise by approximately 3%. One of the employment targets is three hundred fifty thousand new jobs in metropolitan areas [20]. The city has specific goals in relation to land and agriculture as well. It aims to meet the provincial targets by possessing arable land and a comprehensive food production capacity. By putting additional measures in place, the overall emissions of significant pollutants will also be contained within the boundaries of the province.

To achieve these objectives, the city will adhere to the directives issued by the provincial party committee, giving special attention to rapid and excellent economic development. The city is dedicated to effectively carrying out these goals by making use of favorable conditions and opportunities [20]. This strategy calls for maximizing available resources, putting forth the greatest amount of effort, and striving for extraordinary results in order to ensure sustained growth and improvement in economic and social metrics.

4.1 Suzhou's Urban Development: Contemporary and Recent Challenges

Suzhou's long-term economic growth and development will depend on its ability to navigate a number of challenging urban challenges. Even with these recent advancements, the foundation is still lacking for a robust and sustained economic recovery. Emerging industry clusters are still too young to become well-established trends, and issues are preventing the change in the economic drivers from continuing [20]. The persistently low market demand necessitates encouraging consumers' and businesses' confidence and expectations.

It is imperative that the city strengthen its capacity to support innovation and change, particularly in terms of creating significant core technologies and increasing the level of technological autonomy. These areas still need to receive more funding and attention in order to encourage growth [20]. In addition, even with progress made in reform and opening-up programs, significant barriers to the real economy's growth continue to exist, with structural contradictions surfacing in particular sectors of the economy.

Public services including employment, education, pensions, and healthcare continue to fall short of the expectations of the general public. There are still gaps in important areas such as childcare, housing, and urban–rural construction. The following areas need to be given special consideration: production of safety, urban operation, and financial and fiscal management [18]. To meet these challenges, Suzhou needs to show that it is more committed to innovation and political structure reform [20]. Developing a reform-oriented mentality and boosting administrative effectiveness will be necessary to overcome these obstacles and achieve long-term, superior development.

Suzhou is presently facing several challenges related to its industrial structure and economic growth. There is a lot of pressure to modernize and transform the city's industries, as many of them are traditional in nature [18]. With a high degree of external dependence and a heavy reliance on trade processing, the industrial level remains at the bottom of the global value chain. The importance of developing the digital economy and technological prowess in addition to moving past traditional manufacturing is highlighted by this situation.

Moreover, there is a glaring disparity between the resources the environment can offer to support economic development and what is necessary. This paradox shows that a more sustainable approach that reconciles ecological preservation with economic expansion is required [26]. Important public service sectors with shortcomings include healthcare and education, which are progressively failing to meet the needs and expectations of the populace.

Notwithstanding these challenges, Suzhou's social governance still requires development. Despite their early successes, the advantages of these innovative reform and opening-up initiatives have progressively faded. It is now essential for the city to focus on increasing the complexity of its social governance systems in order to manage these evolving issues. This means improving service delivery, optimizing administrative procedures, and ensuring that policies and reforms account for the changing needs of the modern population and the shifting demands of the global economy [18]. Suzhou's future will primarily require resolving these systemic and structural issues to build a more complex, resilient, and sustainable urban environment (Fig. 15).

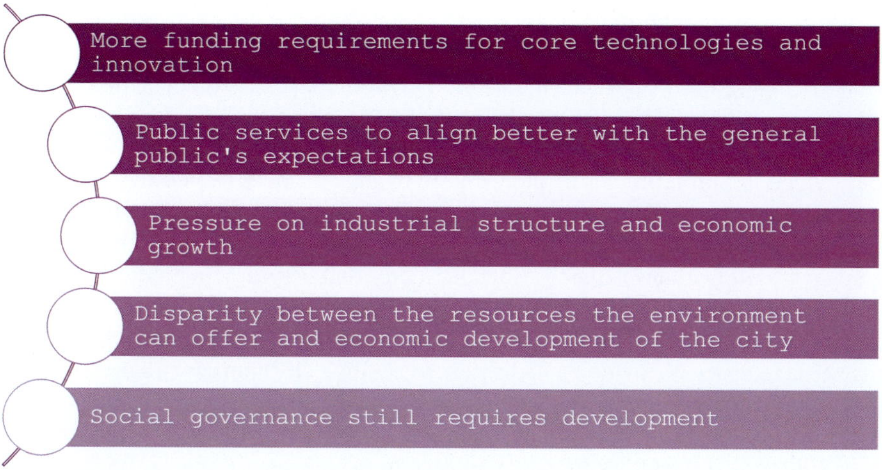

Fig. 15 Summary of main recent and current challenges in Suzhou, which may seem not so unique to Suzhou alone but most of the cities studied in this book

5 Key Lessons Learnt from Suzhou for Other Cities

Cities can learn a lot from Suzhou's development when attempting to strike a balance between modern growth and historical preservation. The city's plan to revive its historic silk industry is an illustration of how preserving traditional craftsmanship can foster both cultural preservation and economic growth. Suzhou's silk industry has thrived because it strongly focuses on quality and legacy while valuing and celebrating its historical roots and contributing significantly to the local economy.

Another key element in Suzhou's success is its Garden Tourism strategy. The city has done a fantastic job of preserving and promoting its classical gardens, which are well-liked tourist attractions as well as significant parts of its cultural legacy. This strategy highlights how important it is to integrate cultural resources into tourism in order to create a sustainable model that enhances visitors' experiences and fosters a sense of community pride. The intentional incorporation of Suzhou into the Yangtze River Delta exemplifies how regional collaboration enhances economic benefits. Suzhou has benefited from being situated in this thriving economic region and has expanded its growth and development through the use of regional networks. This regional integration offers lessons for other cities hoping to take advantage of cooperative opportunities and drive economic advancement through strategic partnerships.

The city's efforts to preserve its canals, which have earned recognition from UNESCO, serve as a reminder of how important it is to preserve historic waterways in the midst of urban development. Suzhou's achievement in receiving this recognition emphasizes how crucial it is to protect historical sites and natural resources while advancing the goals of modern urbanization. This approach can serve as a model for cities attempting to strike a balance between development and preservation. Suzhou's development of sustainable industrial parks offers still another crucial lesson. By highlighting high-tech industries and promoting technological innovation within these parks, the city has attracted substantial investment. Suzhou is a prime example of how innovative industries can grow in well-planned, sustainable industrial zones, which in turn promotes economic growth and advances technology.

Suzhou's heritage district zoning plan serves as an illustration of how historical areas can be enhanced and preserved while combining modern conveniences through the application of urban planning. Suzhou, a city that serves as a model for other cities attempting to strike a balance between historical preservation and contemporary demands, has effectively managed zoning regulations in order to preserve the charm of its heritage districts. Suzhou's dedication to ecologically friendly urban design is a reflection of the city's overall commitment to sustainability. Urban development projects can become more sustainable and liveable by integrating green spaces and eco-friendly practices. Other cities could learn from Suzhou's emphasis on creating green urban areas that enhance quality of life while tackling environmental issues.

Overall, Suzhou's experiences provide a thorough framework that other cities can adopt. Cities can create dynamic and resilient urban environments that honour their heritage while embracing future opportunities by focusing on sustainability, leveraging regional collaborations, enhancing cultural tourism, and blending historic preservation with modern development. Taking everything into account, Suzhou's experiences offer a comprehensive framework that other cities can use. By emphasizing sustainability, utilizing regional partnerships, fusing historical preservation with contemporary development, and promoting cultural tourism, cities can build resilient and dynamic urban environments that honour their past while grasping new opportunities.

References

Bianco L (2001) The Chinese revolution. By Edward J. Lazzerini. [Westport, CT: Greenwood Press, 1999. xx+ 198 pp. ISBN 0–313–30110–7.]. China Q 165:198–198

Cartier C (2021) Exemplary cities in China: the capitalist aesthetic and the loss of space. Urban China reframed. Routledge, New York, pp 17–40

Chang PH (1995) A strategy of memory: Suzhou, China. Doctoral dissertation, Massachusetts Institute of Technology

Chen W, Cheshmehzangi A, Mangi E, Heath T (2022) Implementations of China's new-type urbanisation: a comparative analysis between targets and practices of key elements' policies. Sustainability 14(10):6341

Cheshmehzangi A (2015) Urban identity as a global phenomenon: hybridity and contextualization of urban identities in the social environment. J Hum Behav Soc Environ 25(5):391–406

Cheshmehzangi A (2020) Towards urban energy management: a brief review of China's recent transition in urban energy demand. Low Carbon Econ 11(01):1

Cheshmehzangi A (2020) Identity of cities and city of identities. Singapore Springer

Cheshmehzangi A, Tang T (2022) Yangtze River Delta City Cluster: from integrated production and services to digital technology innovation. In: China's city cluster development in the race to carbon neutrality. Springer Nature Singapore, Singapore, pp 47–62

Cheshmehzangi A, Tang T (2022) Central plains megalopolis: a hub for green transformation and development beyond just the Beijing-Guangzhou channel. In: China's city cluster development in the race to carbon neutrality. Springer Nature Singapore, Singapore, pp 147–159

Duan X, Yu X, Nipper J (2009) Economic polarized trends, function and expanded boundaries of the Yangtze Delta Region. J Geog Sci 19:733–749

Feng Z (2020) The ever-changing technology and significance of silk on the Silk Road. In: The Oxford handbook of history and material culture, p 222

Gao L (2010) 'Breaking and Repairing': conflicting values in the historic gardens of China. Doctoral dissertation, University of Sheffield

Huang L (2000) Housing development in the context of the modernization, urbanization and conservation of Chinese traditional cities: Beijing, Shanghai and Suzhou. Harvard University, Cambridge, MA

Li W, Wang X (2019) Towards a mega-regional economy: a financial geography perspective to understanding Shanghai's role in the Yangtze River Delta integration. Geoforum 101:1–9

Liu B (2023) Human settlement, inhabitation, and travel environment studies in water-net region. In The trialism and application of human settlement, inhabitation and travel environment studies: applications in water-net region. Springer Nature Singapore, Singapore, pp 269–344

Oakes T (2001) Tourism and local economic development in China: case studies of Guilin, Suzhou and Beidaihe. By Xu Gang [Richmond, Surrey: Curzon (English-language series of the Institute of Asian Affairs, Hamburg, No. 4), 1999. 244 pp. ISBN 0–7007–1152-X.]. China Q 165:202–203

Suzhou Industrial Park Management Committee (n.d.) Introduction to Suzhou Industrial Park <<苏州工业园区简介>>. Available from: https://www.suzhouhui.com/point/3512.html

Suzhou Municipal People's Government (2021) The 14th National Economic and Social Development of Suzhou City Five-year plan and outline of long-term goals for 2035. <<苏州市国民经济和社会发展第十四个五年规划和二〇三五年远景目标纲要>>. Available from: https://www.suzhou.gov.cn/szsrmzf/zfwj/202103/0c7ae53f37c94572bb75ce3c888b684d/files/5d664161a6c0436baa8f4c6e8f53d4a5.pdf

Suzhou Municipal People's Government (2022) Suzhou has increased its GDP by one trillion yuan in ten years. <<苏州, 十年增加一万亿GDP>>. Available from: http://www.suzhou.gov.cn/szsrmzf/szyw/202210/55cbdb73ad1f4e008abe9b803c074c9c.shtml

Suzhou Municipal People's Government (2024) 2024 Government work report <<2024年政府工作报告>>. Available from: http://www.suzhou.gov.cn/szsrmzf/zfgzbg/202402/aa953ce81d57475f943b2d27201665df.shtml

Tang B, Cheung P (2020) Suzhou in transition: an introduction. In: Suzhou in transition. Routledge, pp 1–12

Tsao JLJ (1992) Remembering Suzhou: urbanism in late imperial China. University of California, Berkeley

Wang S, Guo Q, Yuan J, Li H, Fu B (2023) Research on the conservation methods of Qu street's living heritage from the perspective of life continuity. Buildings 13(6):1562

Winnicki KM (2021) Imagined cityscapes: new perspectives on future urbanization in contemporary speculative fiction. Doctoral dissertation, State University of New York at Buffalo

Xie L, Cheshmehzangi A, Tan-Mullins M, Flynn A, Heath T (2020) Urban entrepreneurialism and sustainable development: A comparative analysis of Chinese eco-developments. J Urban Technol 27(1):3–26

Xu Y (2015) A cluster analysis comparison of classical Chinese gardens with modern Chinese gardens. Michigan State University, East Lansing

Yang H (2022) Critique and discussion on some typical gardens. In: A treatise on the garden of Jiangnan: a study on the art of Chinese classical garden. Springer Nature Singapore, Singapore, pp 551–673

Zhang S, Zhao J, Jiang Y, Cheshmehzangi A, Zhou W (2023) Assessing the rural–urban transition of China during 1980–2020 from a coordination perspective. Land 12(6):1175

Zhang R, Wang J, Brown S (2021) 'The Charm of a Thousand Years': exploring tourists' perspectives of the 'culture-nature value' of the Humble Administrator's Garden, Suzhou, China. Landscape Res 46(8):1071–1088

Chapter 12
Knowledge Share Between Cities and Progression on China's 30 Years of Urban Change

1 An Insight into China's 30 Years of (Rapid) Urban Change

Between ourselves, we debated whether this book needed a concluding chapter, but ultimately decided that other cities across the world would benefit more from highlighting the lessons China's 30 years of transformation had to teach us. We see this book as a continuation of our first attempt at the city cluster level [7]. Nonetheless, lessons learnt here are especially relevant to developing-nation cities, where the challenges posed by urbanization persist. China has experienced rapid economic growth and urbanization due to a combination of innovative thinking, strategic planning, and responsiveness to regional and global challenges. These experiences offer other cities a wealth of information, including workable solutions and motivational case studies for overcoming the complex issues associated with urban development.

A key takeaway from China's urban development experience is the significance of striking a balance between modernization and heritage preservation [9]. Cities such as Xiamen, Qingdao, Chongqing, and many more (see our volume 2 for reference), have shown that it is feasible to pursue modern development objectives while respecting and integrating historical and cultural assets. This harmony strengthens a city's distinct identity and gives its citizens a feeling of continuity and community [4]. Adopting this strategy can assist developing cities in protecting their cultural legacy while allowing for expansion and change, which will ultimately result in a more lively and cohesive urban setting.

The significance of thoughtful urban planning and infrastructure development is emphasized by the experiences of cities like Shanghai and Shenzhen [11]. China's emphasis on creating cutting-edge, productive infrastructure and encouraging innovation has been essential in enabling the country's quick urbanization and economic expansion. Emerging cities should take note of this and prioritize infrastructure expenditures that boost quality of life, encourage sustainable growth, and improve connectivity (Fig. 1). Prioritizing green infrastructure and smart city technologies can

A. Cheshmehzangi and T. Tang, *30 Years of Urban Change in China's 10 Core Cities*, Urban Sustainability, https://doi.org/10.1007/978-981-97-8846-0_12

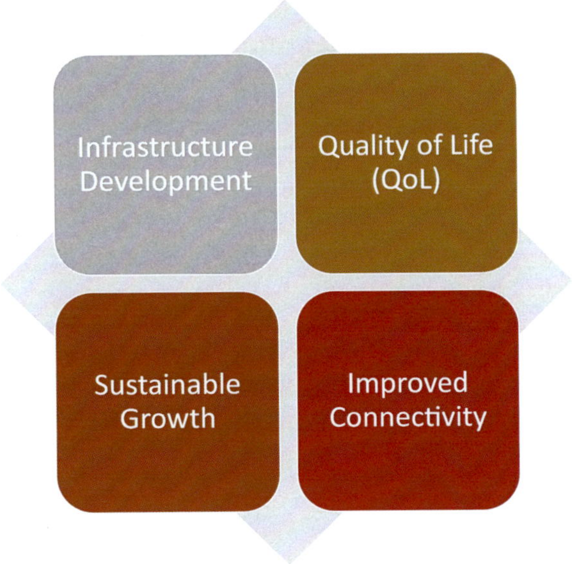

Fig. 1 Suggested priorities for emerging cities through the process of urban change

also aid in addressing urgent issues [13] like resource management and environmental degradation.

Moreover, Chinese cities looking to expand internationally can learn a lot from the country's policies regarding regional integration and international connectivity [8]. Cities can increase their economic and cultural influence by utilizing strategic locations and encouraging cross-border collaborations. In order to draw talent, investment, and cross-cultural exchange, developing cities should think about how they can strengthen their own ties to other countries [3] and participate in international networks.

Finally, Beijing and Chengdu's emphasis on innovation and creative spaces shows how crucial creativity is to urban development. By integrating artistic and cultural venues into urban planning, cities can develop a thriving cultural scene that promotes local artists and attracts tourists. Investing in cultural infrastructure and fostering a creative economy can both accelerate economic growth and improve living conditions in developing cities. In summary, developing-nation cities can benefit from the insights gleaned from China's thirty years of urban transformation. These cities can negotiate the challenges of urbanization and create resilient, dynamic, and sustainable urban environments by emphasizing innovation, cultural tourism, strategic planning, international connectivity, and heritage preservation [1]. Not only do the experiences and tactics covered in this book offer insight into China's achievements [6], but they also serve as a manual for other cities hoping to accomplish similar transformative objectives.

2 Multiple Factors that Matter: From Sustainability to High-Quality Urbanisation and Human-Centric Development Pathways

China's urban transformation over the last three decades has produced a wealth of lessons and best practices that apply to Chinese cities as well as urban areas across the globe. This era of rapid development highlights several essential elements that support successful urbanization, including human-centered design, sustainable development, and high-quality development (Fig. 2). Cities can navigate their own urban opportunities and challenges by knowing these factors and sharing them with one another.

2.1 Pathways to Sustainable Development

In China, sustainability has become a fundamental aspect of urban development and is an essential component for all cities worldwide. China's fast urbanization has

Fig. 2 Summary of multiple factors that matter for China's current phase of urbanization and urban change

unavoidably brought about environmental problems like pollution, habitat destruction, and resource depletion. Chinese cities have embraced sustainable practices more and more in response. To lessen their impact on the environment, for example, Beijing and Shenzhen have made significant investments in renewable energy and green infrastructure projects [5]. The conversion of Shenzhen into a preeminent green city serves as an example of how metropolitan areas can incorporate sustainability into their development strategies. The city has developed energy-efficient buildings, prioritized electric vehicles, and established vast networks of public transportation. In a similar vein, Beijing's attempts to enhance air quality via more stringent emissions regulations and green building guidelines demonstrate a dedication to resolving environmental issues alongside ongoing economic expansion.

It is essential for other cities, particularly those in developing nations, to implement sustainable practices. This entails encouraging energy efficiency, incorporating green technologies, and making sure that environmental health is not jeopardized by urban growth. Cities can lower their carbon footprint, improve the standard of living for their citizens, and support international environmental goals by putting sustainability first.

Superior urbanization goes beyond simple growth to address the general welfare of urban areas. It includes the standard of living conditions, services, and infrastructure. China's experience shows how important it is to improve city livability in addition to city expansion. For example, the Pudong New Area development in Shanghai is an example of a high-quality urbanization commitment. Once a desolate neighborhood, this area is now a thriving financial and commercial center with first-rate public spaces, infrastructure, and amenities. The focus on developing well-thought-out mixed-use developments has increased livability and accessibility, establishing a benchmark for other cities.

2.2 *High-Quality Urbanization*

High-quality urbanization in developing cities entails prioritizing the creation of environments that are safe, secure, and appealing for inhabitants. It entails making investments in inclusive communities, guaranteeing the provision of basic services, and comprehensive urban planning. By doing so, the dangers of haphazard growth are avoided, and more resilient and sustainable urban environments are produced. Another important element that has been essential to China's urban evolution is human-centric development. This strategy puts residents' needs and wellbeing first, emphasizing the creation of inclusive and equitable urban environments. It entails dealing with problems like community involvement, affordable housing, and accessibility to amenities. One example of human-centric urban planning is the growth of Chengdu's Tianfu New Area. The city has placed a strong emphasis on developing residential areas with easily accessible green areas, first-rate healthcare facilities, and educational opportunities. By putting residents' needs first, the overall quality of life has been improved by fostering a sense of community and belonging.

2.3 *Human-Centric Approach to Urban Development*

Adopting a human-centric approach for cities around the world entails talking to locals to learn about their needs and including their input in planning procedures. Ensuring that all citizens can profit from urban development, it also entails creating inclusive, accessible, and reasonably priced urban environments. This strategy improves urban residents' quality of life and social cohesiveness. The Chinese urban change trajectory highlights the importance of knowledge exchange between cities. Through assimilating the achievements and obstacles encountered by Chinese cities, other metropolitan areas can modify and execute efficacious tactics customized to their distinct circumstances. Sharing knowledge entails not just copying effective models but also comprehending the underlying ideas and modifying them for regional circumstances. Shenzhen's green initiatives, for instance, are noteworthy, but other cities may need to modify these tactics to suit their unique economic and environmental circumstances. Similar to this, excellent urbanization techniques from Shanghai can be modified to meet the requirements of various urban environments while taking into account the limitations and resources of the area.

Innovation and progress can be accelerated by cooperation and exchange between cities. Cities with comparable problems can gain from each other's experiences and perspectives, which can lead to an international conversation about best practices and solutions. Platforms for this exchange are provided by conferences, international partnerships, and urban networks, which enable cities to build on one another's successes and propel collective advancement.

China's 30 years of urban development have yielded a wealth of information about human-centered development, high-quality urbanization, and sustainability [6, 10, 12, 14, 15]. These elements are necessary to create inclusive, livable, and resilient cities. Cities all over the world can better address their own urbanization challenges and build urban environments that promote the well-being of all citizens by exchanging these insights and tailoring them to local contexts [2]. The knowledge gained from China's experience will continue to be an important tool in determining how cities around the world will develop in the future.

3 Further Discussions: Lessons from Other Cities for Chinese Cities

China's rapid urbanization and transformation make it worthwhile to look outside its borders and take note of what other cities around the world have learned. Chinese cities can learn valuable lessons from the successes and difficulties faced by cities in various contexts, which they can apply to tackle their own urban development issues. Here, we examine the most important lessons that Chinese cities can learn from the urban experiences around the world, with a particular emphasis on issues like

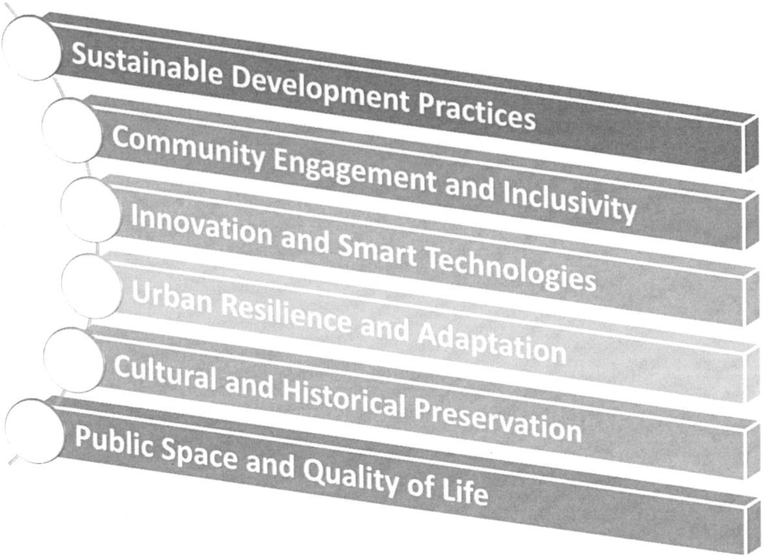

Fig. 3 Summary of six key lessons or knowledge shared from other cities for Chinese cities

innovation, sustainable development, community involvement, and urban resilience (see Fig. 3).

3.1 Sustainable Development Practices

Sustainability is becoming a top priority for cities all over the world to improve livability and address environmental issues. For instance, Chinese cities can learn a lot from Copenhagen's pledge to become carbon–neutral by 2025. This is a similar transformative change that many Chinese cities are going through the expected sustainability transitions. For instance, Copenhagen's urban planning has incorporated green infrastructure, like vast bicycle networks and energy-efficient buildings. In addition to enhancing environmental quality, this emphasis on sustainability has increased the city's resilience and allure. Cities in China stand to gain from implementing comparable sustainable practices. Mitigating environmental impacts can be achieved through integrating green spaces, promoting energy-efficient building standards, and improving public transportation. Moreover, Chinese cities hoping to lower pollution and traffic congestion could take a cue from cities like Amsterdam, which have shown the advantages of investing in cycling infrastructure. Urban environments that are healthier and more resilient can be achieved by prioritizing sustainability in urban planning.

3.2 Community Engagement and Inclusivity

Inclusion and community involvement are essential elements of successful urban development. Cities that involve their citizens in decision-making, like Portland, Oregon, have embraced participatory planning techniques. Portland's strategy makes sure that a variety of perspectives are heard and taken into account in urban development by utilizing community forums, advisory committees, and interactive planning tools. Promoting community involvement in Chinese cities can increase the significance and efficacy of urban projects. By involving locals in the planning process, needs and concerns are addressed, resulting in more equitable and inclusive urban environments. Similar community workshops and public consultations are examples of participatory strategies that Chinese cities can use to engage locals in creating their urban environments. Social cohesion can be strengthened and life quality raised as a result.

3.3 Innovation and Smart Technologies

Global urban landscapes are changing because of innovation and the application of smart technologies. With the use of technologies like data-driven urban planning, intelligent lighting, and sophisticated traffic management systems, cities like Singapore have emerged as leaders in the field of smart city solutions. Through the use of technology, Singapore's Smart Nation initiative seeks to improve urban living by increasing the city's efficiency and responsiveness to its citizens' needs. Chinese cities can incorporate smart technologies into their infrastructure and urban planning, taking a cue from Singapore's approach. Energy-efficient systems, digital public services, and smart traffic management are just a few examples of innovations that can increase city productivity and citizen satisfaction. Using technology to solve problems in cities also makes cities leaders in the global smart city movement.

3.4 Urban Resilience and Adaptation

For cities to withstand the effects of shocks like climate change, urban resilience is essential. The rehabilitation of New York City following Hurricane Sandy offers an insightful case study for developing urban resilience. To better withstand unforeseen events, the city has put in place safeguards against flooding, enhanced emergency response systems, and resilient infrastructure. Similar resilience strategies can be adopted by Chinese cities, many of which are susceptible to environmental challenges and natural disasters. Chinese cities can become more resilient by investing in climate adaptation measures, creating comprehensive disaster response plans, and constructing infrastructure that can withstand extreme weather events. Building more resilient urban environments that can adjust to new challenges can be facilitated by taking lessons from New York's experiences.

3.5 Cultural and Historical Preservation

In the face of rapid urbanization, preserving a city's cultural and historical identity is crucial. Paris and other cities have been able to successfully strike a balance between modernization and the preservation of their rich cultural legacy. Paris's strategy entails safeguarding historical locations, tastefully incorporating contemporary construction, and encouraging cultural tourism. Cities in China, with their rich historical heritage and dynamic terrain, stand to gain from implementing comparable tactics. Historic sites can be preserved and included in urban development plans to support growth and preserve cultural heritage. With their rich histories, cities like Suzhou and Xi'an can learn from Paris's example to make sure modernization does not compromise their cultural identity.

3.6 Public Space and Quality of Life

Urban living is greatly impacted by the quality of public spaces. High-quality public areas, such as parks, plazas, and waterfronts, are characteristic of cities like Melbourne and enhance social interaction and general well-being among locals. Melbourne's efforts to create aesthetically pleasing and easily accessible public spaces improve the city's liveability and encourage community involvement. China's cities can benefit from Melbourne's focus on designing public spaces. Encouraging public spaces that are inclusive, safe, and well-designed can enhance the standard of living for locals. Improving public spaces promotes community development and social interaction while also improving the general aesthetics and functionality of cities.

In conclusion, Chinese urban centers can more successfully handle their development challenges by taking a cue from other cities' experiences. Global insights can improve and inform China's urban strategies in several areas, including high-quality public spaces, cultural preservation, innovation, resilience, sustainable practices, and community engagement. By embracing these insights, Chinese cities will be better equipped to build dynamic, inclusive, and resilient urban environments that serve their citizens' needs and advance international trends in urban development.

4 A Brief SWOT Analysis of China's Urban Change in Recent 30 Years

4.1 Strengths

Over the past three decades, China's urban transformation has been characterized by significant infrastructure development and rapid economic growth. The speed and scope of urban expansion, which has turned many cities into major international economic centers, is one of its biggest advantages. Large cities like Beijing, Shenzhen, and Shanghai have developed into global hubs for innovation, technology, and finance. Urban efficiency and connectivity have increased with the development of modern infrastructure, such as extensive public transportation networks, high-speed rail, and smart city technologies. Furthermore, China has established its cities as leaders in intelligent and environmentally friendly urban solutions thanks to its proactive approach to incorporating innovation and technology into urban planning.

4.2 Weaknesses

China's urban development exhibits notable weaknesses despite its impressive growth. Growing urbanization has put a strain on natural resources and created serious environmental problems, such as pollution of the air and water. Furthermore, the rapid growth has occasionally left some new districts with inadequate infrastructure and poorly planned urban areas. Socioeconomic inequality and housing affordability have also become issues, with some locals having trouble obtaining decent housing and public services. Concerns regarding social justice and community cohesion have also been raised by the widespread uprooting of communities brought about by redevelopment projects.

4.3 Opportunities

The urbanization of China offers numerous chances for advancement and development. There is a chance to build more resilient and environmentally friendly urban environments with an ongoing focus on sustainable development and green technologies. Additionally, there's a chance to use smart city technology developments to improve quality of life and urban management. Incorporating cultural and historical preservation into urban planning can also promote a distinct urban identity and draw tourists. With the increasing development of China's urban areas comes the chance to concentrate on high-quality urbanization that better serves the needs of the populace and fosters social justice and inclusivity.

4.4 *Threats*

The course of China's urban development in the future may be affected by several threats. The sustainability of urban growth is significantly threatened by environmental degradation and climate change, which could worsen pollution and resource scarcity. The growth and stability of urban economies could be impacted by trade tensions and economic fluctuations, especially in cities that depend heavily on exports. Furthermore, social problems like rising inequality and housing affordability concerns may cause social unrest and discontent among locals. If not carefully managed, the potential for rapid, unplanned urban expansion could put a strain on services and infrastructure, causing issues with quality of life and urban management.

5 Lessons from China's Urban Change: Looking Back, Looking Forward

Given the rapid global urbanization that has occurred over the past three decades, China's urban transformation provides a wealth of lessons for cities around the world. There are important lessons to be learned from China's rapid urban development, lessons that can inform future urban planning and policy.

5.1 *Looking Back: Key Achievements and Challenges*

China's urbanization has seen unheard-of levels of expansion and development. Large cities with cutting-edge infrastructure and thriving economies, such as Shanghai, Beijing, and Shenzhen, have developed into global economic powerhouses. Significant infrastructure investments, such as high-speed rail, smart city technologies, and contemporary public transportation networks, were the driving forces behind this quick expansion. These developments have raised China's cities to the forefront of technological innovation by greatly improving urban efficiency and connectivity.

The rapid pace of urbanization has not, however, been without difficulties. The state of the environment has deteriorated significantly, with many urban areas experiencing critical levels of air and water pollution. The capacity for efficient planning was frequently outpaced by the quick speed of development, which resulted in problems like uneven urban growth and poor infrastructure in newly developed areas. Furthermore, socioeconomic gaps have grown wider, making access to high-quality services and affordable housing urgent concerns. Social cohesion and equity have become more pressing concerns as a result of communities being uprooted for redevelopment.

5.2 *Looking Forward: Opportunities and Strategic Directions*

China can advance in multiple strategic directions that leverage previous learnings to promote more inclusive and sustainable urban development. Highlighting eco-friendly technologies and sustainability is still a vital opportunity. Chinese cities can lessen the effects of the environment and improve urban resilience by putting a strong emphasis on energy efficiency, pollution control, and environmentally friendly infrastructure. The creation of green areas, sustainable transportation networks, and renewable energy sources will be essential to reaching these objectives. The incorporation of smart city technologies is another essential area for future development. Utilizing developments in digital infrastructure, artificial intelligence, and data management can boost service delivery, optimize urban management, and boost productivity all around. By utilizing technology, cities can make their urban environments more responsive and adaptable by addressing problems like waste management, traffic congestion, and energy consumption.

Furthermore, the importance of human-centric development cannot be overstated. Fostering social cohesion and raising living standards will require ensuring inclusive and equitable urban growth. In order to guarantee that communities' needs and concerns are taken into account in development plans, this entails addressing housing affordability, improving access to necessary services, and involving communities in the planning process. Another big opportunity is to preserve cultural and historical heritage in the face of fast modernization. Cities can retain their distinctive identities and draw tourists by incorporating historical preservation into their urban planning, which enhances both the cultural diversity and the vibrancy of the economy.

Lastly, China's urban transformation provides valuable insights into urban contexts in the country and around the world. The triumphs and obstacles encountered in the last three decades offer significant perspectives on the intricacies of swift urbanization. Looking ahead, creating resilient, inclusive, and dynamic urban environments will require a focus on sustainability, smart technologies, human-centric development, and heritage preservation. Chinese cities have the potential to maintain their leadership position in innovative urban development and to create sustainable and equitable urban futures by strategically addressing emerging challenges and reflecting on past experiences.

References

Bai X, McPhearson T, Cleugh H, Nagendra H, Tong X, Zhu T, Zhu YG (2017) Linking urbanization and the environment: conceptual and empirical advances. Annu Rev Environ Resour 42(1):215–240

Cappa F, Franco S, Rosso F (2022) Citizens and cities: leveraging citizen science and big data for sustainable urban development. Bus Strateg Environ 31(2):648–667

Castells M (1983) The city and the grassroots: A cross-cultural theory of urban social movements (No. 7). University of California Press

Cheshmehzangi A (2020) Identity of cities and city of identities. Springer, Singapore

Cheshmehzangi A (ed) (2021) Green infrastructure in Chinese cities. Springer, Singapore
Cheshmehzangi A, Dawodu A, Sharifi A (2021) Sustainable urbanism in China. Routledge, New York
Cheshmehzangi A, Tang T (2022) China's carbon neutrality plan implementation and sustainable development process review. In: China's city cluster development in the race to carbon neutrality. Springer Nature Singapore, Singapore, pp 1–23
Dameri RP, Benevolo C, Veglianti E, Li Y (2019) Understanding smart cities as a glocal strategy: a comparison between Italy and China. Technol Forecast Soc Chang 142:26–41
Demgenski P (2015) Seeking a future for the past: negotiating inner city redevelopment and heritage in Qingdao. The Chinese University of Hong Kong (Hong Kong), China
Fang C, Yu D, Fang C, Yu D (2016) The developmental strategies and basic principles for China's new urbanization. In: China's new urbanization: developmental paths, blueprints and patterns, pp 49–109
Lin W, Ghose R (2010) Social constructions of GIS in China's changing urban governance: The case of Shenzhen. Cartographica: The Int J Geogr Inf Geovisualization 45(2):89–102
Tillu J (2021) The economic case for China's sustainable urban transformation. The London School of Economics and Political Science. Available from: https://www.lse.ac.uk/granthaminstitute/wp-content/uploads/2021/07/Economic-case-for-Chinas-sustainable-urban-transformation.pdf
Xie L, Tan-Mullins M, Cheshmehzangi A (2019) Political ecology of Chinese smart eco-cities. In: Zhang, X. (ed.) Remaking sustainable urbanism. Palgrave Macmillan, Singapore. https://doi.org/10.1007/978-981-13-3350-7_4
Zhang L, LeGates R, Zhao M (2016) Understanding China's urbanization: the great demographic, spatial, economic, and social transformation. Edward Elgar Publishing, Cheltenham
Zhang S, Zhao J, Jiang Y, Cheshmehzangi A, Zhou W (2023) Assessing the rural–urban transition of China during 1980–2020 from a coordination perspective. Land 12(6):1175

SPRINGER NATURE

GPSR Compliance

The European Union's (EU) General Product Safety Regulation (GPSR) is a set of rules that requires consumer products to be safe and our obligations to ensure this.

If you have any concerns about our products, you can contact us on ProductSafety@springernature.com

In case Publisher is established outside the EU, the EU authorized representative is:

Springer Nature Customer Service Center GmbH
Europaplatz 3
69115 Heidelberg, Germany

The manufacturer's authorised representative in the EU is Springer Nature Customer Service Centre GmbH, Europaplatz 3, 69115 Heidelberg, Germany. If you have any concerns regarding our products, please contact ProductSafety@springernature.com

Printed and bound by CPI Group (UK) Ltd, Croydon, CR0 4YY
13/10/2026
02232321-0001